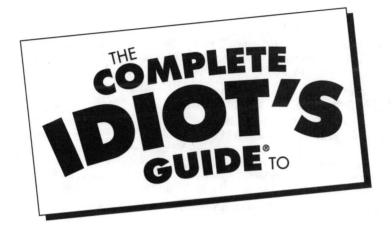

THE
COMPLETE
IDIOT'S
GUIDE® TO

Understanding the Brain

By Arthur Bard, M.D. and Mitchell Bard, Ph.D.

ALPHA

Mitchell Bard would like to dedicate this book to his mother who finally gets to hear her son sound like a real doctor. Arthur Bard would like to dedicate the book to his long suffering wife, and mother of the coauthor, for putting up with him during the ordeal of writing this book.

International Standard Book Number: 0-02-864310-0
Library of Congress Catalog Card Number: 2002106349

05   04   03   8   7   6   5   4   3   2

Interpretation of the printing code: The rightmost number of the first series of numbers is the year of the book's printing; the rightmost number of the second series of numbers is the number of the book's printing. For example, a printing code of 02-1 shows that the first printing occurred in 2002.

Printed in the United States of America

Note: This publication contains the opinions and ideas of its author. It is intended to provide helpful and informative material on the subject matter covered. It is sold with the understanding that the author and publisher are not engaged in rendering professional services in the book. If the reader requires personal assistance or advice, a competent professional should be consulted.

The author and publisher specifically disclaim any responsibility for any liability, loss, or risk, personal or otherwise, which is incurred as a consequence, directly or indirectly, of the use and application of any of the contents of this book.

Publisher: *Marie Butler-Knight*
Product Manager: *Phil Kitchel*
Managing Editor: *Jennifer Chisholm*
Acquisitions Editor: *Randy Ladenheim-Gil*
Development Editor: *Suzanne LeVert*
Production Editor: *Billy Fields*
Copy Editor: *Heather Stith*
Illustrator: *Chris Eliopoulos*
Cover/Book Designer: *Trina Wurst*
Indexer: *Angie Bess*
Layout/Proofreading: *Megan Douglass, Becky Harmon, Mary Hunt, Sherry Taggart*

Contents at a Glance

Contents

Introduction

This book was not an easy one to write. The brain surgeon of the two of us (Arthur Bard, M.D.) thought the entire project might be impossible because it was difficult for him to conceive reducing the complexity of his life's work to the format used in this series.

We found ourselves trying to strike a delicate balance between being informative and comprehensive and being interesting and understandable. In looking at other books written on the brain, we found they usually were on one end of the spectrum or the other.

Many books are written for children on the human body; these books are useful in that they simplify the material, but they are unsatisfactory because whole topics are typically reduced to a paragraph. On the other hand, we didn't want this book to be a text for medical students either, loaded with jargon and discussions of every structure and function of the brain.

We hope that we have found the right balance and offer you an opportunity to learn not only the basics about the brain, but also some of the more sophisticated concepts without making you feel overwhelmed. Medical research makes new advances every day, and we've tried to introduce some of the latest findings in addition to the more established ideas about the brain.

Those of you who are familiar with this series may notice that the tone shifts in the latter half of this book. This is quite deliberate. Although we have adopted the normal light-hearted approach for much of the book, we also felt the seriousness of the chapters relating to disease, drug abuse, and treatment required a more businesslike style. We are confident that you will find all of the material we offer interesting and useful nonetheless.

We don't expect you to be ready to perform brain surgery after you read *The Complete Idiot's Guide to Understanding the Brain*, but we do hope that you'll have a greater appreciation for the amazing and mysterious organ that is allowing you to read and comprehend these words.

What You'll Find in This Book

Part 1, "The First Brain," introduces you to the mysteries of the brain and highlights how the brain has been viewed by philosophers, scientists, theologians, and physicians through the centuries. It also tracks the evolution in the understanding of the structure and function of the brain.

Part 2, "Anatomy 101," gives you the nitty-gritty on the structure of the brain. You'll learn all about the different parts of the brain and what they do.

Part 3, "The Human Computer," focuses on many of the key functions of the brain, such as providing the ability to speak, see, hear, touch, taste, and smell. This part looks at basic needs, including eating, drinking, sleeping, and engaging in sexual activity. Finally, this part examines some of the involuntary and voluntary actions that the brain controls.

Part 4, "Acting Out," provides insights into memory, intelligence, and emotion.

Part 5, "The Sick Brain," describes a wide variety of injuries, diseases, and disorders affecting the brain. The chapters in this part describe everything from headaches to brain tumors to strokes. This part also covers mental illness and the effects of drug and alcohol abuse on the brain.

Part 6, "Treatment (Couches, Shocks, Pills, and the Knife)," describes the various ways that problems related to the brain can be diagnosed, treated, and sometimes cured.

Extras

Extras offer little bits of information and insight sure to whet your appetite when it comes to learning more about the brain. They are presented as boxed sidebars to draw your attention.

Words of Wisdom

These sidebars offer definitions to some of the difficult terms you'll find in the text.

IQ Points

These sidebars offer fascinating glimpses into the way the human brain works.

Code Blue

These words of caution concern things that can happen in the brain or warnings about matters relating to how the brain is understood.

Gray Matter

Here, you'll find additional items of interest related to the study of the brain.

Acknowledgments

Dr. Arthur Bard would like to thank his early teachers, Professors Eben Alexander and Courtland Davis who piqued his interest in neurosurgery. However, it was not until he met neurosurgeon Fred Rehfeld in Fort Worth, Texas while serving in the Public Health Service that he made the decision to pursue a career in neurosurgery. Doctors Keasley Welch, Richard Lende, Wolf Kirsch, and Tom Craigmile honed his skills as a neurosurgeon during his residency training at the University of Colorado Medical Center.

Dr. Mitchell Bard would like to thank his family for their patience during the often time-consuming process of writing this book. He also wants to thank his father for his patience in dealing with someone who wanted everything explained to him like he was a six-year-old.

Both authors would like to thank Randy Ladenheim-Gil for her interest in publishing such a difficult book and commitment to making it a success. We also are very appreciative of the hard work put into the project by our development editor Suzanne LeVert who had the difficult task of sometimes translating our inscrutable language into readable thoughts.

Trademarks

All terms mentioned in this book that are known to be or are suspected of being trademarks or service marks have been appropriately capitalized. Alpha Books and Penguin Group (USA) Inc. cannot attest to the accuracy of this information. Use of a term in this book should not be regarded as affecting the validity of any trademark or service mark.

Part 1

The First Brain

This part of the book looks at the insightful and sometimes bizarre notions philosophers, physicians, and the common folk have had about the brain through the ages. If you don't have much interest in history, feel free to jump ahead to Part 2. But if you like reading about cool stuff like mummies, hypnotism, and Frankenstein, stick around and you'll learn some interesting things about the way the understanding of the brain has evolved from ancient times to the present.

It All Starts Here

In This Chapter

- ◆ The universe's most complex object
- ◆ The brain's evolution
- ◆ Hard wiring
- ◆ Brain versus computer
- ◆ Brain experts and what they do

Why is the brain so fascinating?

It determines what we think and how we interpret our world. It produces our dreams and nightmares. It tells us to be happy or sad. We eat, drink, and engage in sexual activity because of instructions from our brains. The brain controls all of our internal processes, such as the beating of our heart, the digestion of our food, and the rate of our breathing, without us having to even think about it.

One reason to study the brain is sheer curiosity. The brain has long held the fascination of scientists and physicians even though it was not considered an important organ for many centuries. More recently, as researchers have learned more about the form and function of the brain, it has become even more interesting to laypeople. By getting a better grasp of what the brain looks like, how it is put together, and how it works, we hope to change our lives and those of our children for the better. We hope to find ways to improve our memories, modify our children's behavior, and overcome the anxieties that hinder our actions and affect our happiness.

In addition, the brain is of acute interest because an estimated 50 million people suffer from neurological diseases. The necessity of finding treatments and cures is a driving force behind much brain research.

In the forthcoming pages, you'll have the opportunity to benefit from this research by increasing your understanding and appreciation of nature's most remarkable creation: the human brain. Relax, enjoy, and don't worry—you won't see a drop of blood!

Nature's Marvel

Your brain weighs less than that of a dolphin. It looks like a walnut, but it is squishy like a sponge because it is almost 80 percent water. For those who have ever been scuba diving or snorkeling and seen "brain coral," the large, roundish organism that is covered with curved ridges that looks like a maze was drawn on it, you have glimpsed what the surface of your brain looks like.

> **IQ Points**
>
> Ever wonder why a cut on your head bleeds profusely while a cut on your knee usually just produces a trickle of blood? The reason is that so much blood is pumped to your head that the scalp has a greater blood supply than any other part of your skin. This large blood supply is why head wounds often look much worse than they really are.

The brain is housed in a custom-size bony protective container. Though it is only 2 percent of your body's weight, it demands 20 percent of your body's fuel. Approximately one fifth of the blood your heart pumps (about 5 quarts per minute) is sent to the brain via your body's four main arteries. The brain essentially runs on just oxygen and glucose and generates about the same amount of energy as a 10-watt light bulb. It works 24/7 for as long as you live, and it has no moving parts.

Put simply, your brain is the most incredible and complex object that exists in the universe.

Beagles and Brains

In 1831, a 22-year-old British naturalist named Charles Darwin joined a scientific expedition aboard the English survey ship HMS *Beagle*. During the around-the-world voyage, Darwin made careful and astute observations about the animals in different regions. Based on the similarities and differences he saw, Darwin developed his theory of evolution. According to Darwin, a species that survives over generations has certain advantages that allow it to compete with other species (or within their own). Through the process of *natural selection*, these advantages are passed on to the next generation. For example, animals that are the best hunters survive, and the characteristics that make them good hunters (for example, keen eyesight, speed, stealth) will be inherited by their offspring. Another key element of his theory, which is particularly relevant to the study of the brain, is the idea that all related organisms are descended from common ancestors.

Darwin's theory was not readily accepted be-cause he couldn't prove it; it still has some doubters to this day. Furthermore, his ideas were viewed as a direct challenge to religious beliefs about the creation of the world by a divine being. According to the creation story in the Bible, God made man roughly 6,000 years ago. In the latter half of the century, however, pale-ontologists, archaeologists, and anthropologists began to discover links between humans and apes. They also uncovered ancient cultures and other evidence that traced the birth of humans to a far earlier period and illustrated an evolu-tionary sequence of development.

Words of Wisdom

Natural selection is a process that results in the survival of plants or animals that are best suited to their environment and leads to the perpetuation of genetic qualities best adapted to that environment.

A Monkey's Uncle

The development of the brain also follows an evolutionary path that begins with the first animals that lived in the sea. Because their lives were relatively simple, these creatures had brains consisting of only a small number of nerve cells.

When the early reptiles emerged from the seas some 300 million years ago, they needed to move and to have keen senses for locating prey. Their brains grew to include the cerebellum and the components of the brain stem, the components that control movement and the senses.

The first mammals appeared approximately 200 million years ago. They maintained the traits of the reptiles, but added additional qualities such as memory and emotion. Thus, their brains de-veloped new structures to meet these needs.

The appearance of the first primates and early *hominids* marked a period of rapid growth in brain development. Fossils show that the human body grew larger over time, but the brain evolved more rapidly.

Less than two million years ago, a greater mem-ory, more fine motor skills, language, and more advanced thought processes were accompanied by the development of the cerebrum. We retained the older parts of the brain from our animal ances-tors, but added more components.

Words of Wisdom

The human family is defined as **hominids**. Character-istics of our family include a large, highly developed brain, our upright position, and our manner of movement. We are also distinguished from other ani-mal families by our construction and use of tools.

The human brain enables our species to survive in its surroundings. If our brains did not evolve with all of these components, animals probably would have killed us off a million years ago, and they would rule the planet.

Human Evolution

Date	Characteristics	Brain size (cubic centimeters)
4–2.75 million years ago	Our most ancient ancestor is *Australopithecus afarensis*. The skeletal remains are known as "Lucy."	380–450 cc
3–1.6 million years ago	*Australopithecus africanus* was taller than "Lucy" and had a greater brain capacity.	380–450 cc
2.3–1.3 million years ago	*Australopithecus robustus* was significantly taller than its predecessors and had a much larger brain.	500–600 cc
1.8–1 million years ago	*Homo erectus* had a similar body size to modern humans. Probably used tools such as handaxes and may have made use of fire and occupied caves.	800–1300 cc
200,000–300,000 years ago	*Homo sapiens* began to look more "human" and less apelike. Their skulls were rounder and larger, and their teeth and jaws smaller than *erectus*.	1350 cc
100,000–40,000 years ago	Neanderthals were short and muscular. The skull of the Neanderthal was long with a low forehead and massive chin. Their brains were large (bigger than ours), but organized differently than modern humans. In particular, they had a smaller neocortex. Nevertheless, the stereotype of Neanderthals as dim-witted brutes is inaccurate.	1500 cc
35,000 years ago	*Cro-magnon* began to show evidence of innovations in art (such as the cave paintings found in France) and technology (e.g., new hunting tools).	1600 cc

Though we look down our noses at our ancestors, the truth is the cavemen were not the dumb brutes we think they were. The brain in your head is essentially the same as the one the people of the Cro-Magnon period had 35,000 years ago. That's right: If you had put a computer in front of that hairy guy in the loincloth carrying a club, he would have had the mental capacity to use it. The problem, of course, was that it took thousands of years to develop the computer technology.

The Baby Brain

The evolutionary process that resulted in our brains looking the way they do today took millions of years, but the development of an individual human brain from conception to birth is a far more rapid affair. Perhaps even more amazing, the entire blueprint for building it is contained on a six-foot-long strip of genetic material known as *DNA*.

For much of the nine months spent in the womb, the newly formed being has no consciousness, knowledge, or feelings.

About 18 days after fertilization, a neural plate forms from the embryonic tissue, and this plate slowly becomes a tube that grows thicker and soon develops three enlargements: the *forebrain*, the *midbrain*, and the *hindbrain*.

By the seventh week of development, these three main structures have divided further, and the more specialized areas of the brain begin to form. The cerebral hemispheres expand from the forebrain to the point where they overshadow all other parts of the brain. The midbrain swells, and four lobes associated with vision and hearing form. A portion of the hindbrain forms into the cerebellum while other growths produce the remainder of the brain stem. The outlines of the brain become apparent by the third fetal month.

Meanwhile, about 250,000 new neurons are being created every minute in the womb; virtually all of the neurons you will have during your lifetime are generated before birth. Neurons are the basic building blocks of the nervous system. They are cells typically consisting of a cell body, axon, and dendrites. Neurons transmit nerve impulses to and from the brain.

Words of Wisdom

DNA is short for deoxyribonucleic acid. The acronym is shorthand for describing the nucleic acids in cells that form the shape of a double helix and are the molecular basis of heredity.

Words of Wisdom

The **forebrain** includes the cerebrum, thalamus, and hypothalamus. The **midbrain** is, surprise, the middle of the brain. To be more specific, it is the uppermost part of the brain stem. The **hindbrain** is the lower region of the brain stem, comprising the pons and medulla.

The parts of the brain.

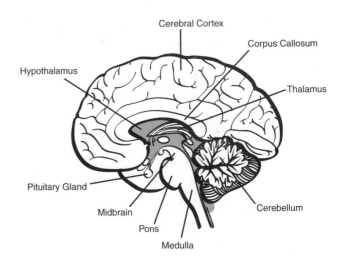

By the time infants are born, their brains and other organs have developed sufficiently for them to survive outside the womb. Yet nature also has made sure the brain remains small enough for a baby's head to pass through the birth canal. If the brain were fully formed, this passage would be impossible because an infant's head would be too big.

You Were Smarter As a Kid

Most of the remainder of brain development occurs in the first two years of life. The brain that started out about 12 ounces at birth rapidly grows to about 35 ounces by the end of the first year. By age five, the brain is roughly the size and weight it will remain throughout a person's life.

Although most of the neurons are in place before birth, they are not wired together. That takes place from birth onward as experiences begin to cause the neurons to fire, and each time they do, new connections are built between them. The ability to learn is most pronounced during these early childhood years. Once the neuronal connections are established, picking up new knowledge starts to become more difficult. Ah, to be a kid again!

You may have noticed that children often learn new things more easily than adults do. This difference in learning is because most of the connections of the brain become permanent by the time we have matured. Although parents sometimes joke about their kids losing

CAUTION

Code Blue

Don't wait until your child is a teenager to introduce him or her to a new language. We learn to speak and understand language during early childhood when the brain's wiring is first being established. The neurons in the part of the brain that controls language become connected by age six. This ongoing neuron connection is one reason why children usually have a much easier time picking up new languages than adults do.

brain cells whenever they bonk their heads, it is the adults who are actually losing their brains, bit by bit. Starting around age 20—that's right, age 20—brain cells begin to die at a rate of 10,000 per day and are never replaced.

Over a lifetime, millions of brain cells are lost. Typically, the ability to learn and other physical and mental attributes begin to deteriorate as a person grows older. You're not washed up at 30, but by the time you're 60, you'll probably begin to notice a difference in what you can and can't do and learn. But by then you'll also have had a long time to accumulate knowledge and experience to compensate.

Assembly Not Required

The brain has often been compared to the most advanced piece of technology available. In the past, for example, the comparison was made to the telephone switchboard. Today, the brain is typically said to resemble a computer, but the way it works is quite different.

Both the computer and your brain require energy to work. The computer's energy comes from electricity, whereas the brain's energy is derived from the food you eat. The computer can be turned off, but the brain is always working.

The brain's functions start with chemical changes that produce electric currents instead of starting with pure electricity like the computer. In addition, the brain's chemical composition is continuously changing whereas the computer's hardware is what it is forever unless you manually upgrade it.

The ongoing chemical changes within the brain mean that it may react differently to the same stimuli; the computer should churn out the same output to an unchanging input.

Microprocessors are becoming increasingly complex more and more information can fit into a smaller and smaller package. For example, a 1 gigahertz microprocessor contains approximately 22 million transistors on a chip the size of a postage stamp. By contrast, a piece of brain the size of a match head could have up to a billion connections. Take that, Bill Gates!

All of those microprocessors enable computers to perform remarkable calculations and numerous tasks at speeds beyond our abilities. At the same time, however, the human body is doing its own multitasking at lightning speeds by maintaining your heart rate and blood pressure and allowing you to carry on a conversation simultaneously.

IQ Points

In May 1997, world chess champion Gary Kasparov was beaten by Big Blue, a chess computer programmed by a team of brilliant chess players, engineers, and computer specialists. Some argued that this victory was a milestone in the development of intelligent computing, while others said it was not the computer that defeated Kasparov, but the team of humans who created it.

If a component of a computer breaks, it usually needs to be replaced or the whole thing may not work. In some instances, when a part of the brain is damaged, it can fix itself! For example, an injury to one area of the brain may result in other parts of the brain taking over its functions.

Individual parts of a computer also work largely independently. Memory is stored in the hard drive. Audio is produced by the sound card. Video is produced by the video card. Though for many years it was believed that specific areas of the brain controlled discrete functions, we now know that this is not true. Several areas of the brain work together at one time to perform a particular function.

The most important difference between a computer and the brain is that, for now at least, the computer does not experience emotions, dreams, thoughts, or awareness of its existence. The human brain does.

Gray Matter

Alan Turing published a paper in 1950, *Computing Machinery and Intelligence*, which suggested that computers would eventually be programmed to acquire abilities rivaling human intelligence. He proposed a test to determine when this programming was successful. In the Turing Test, a human and a computer are questioned by someone who would receive the answers without knowing whether they came from the computer or the person. Turing argued that if the interrogator could not tell who the human was by simply looking at the answers, then the computer could be considered intelligent.

Perhaps a better analogy is to liken the brain to the conductor of an orchestra. Many different instruments make up the orchestra, all with different sounds and structures. They can act independently, but if they do, the result is often chaos. The conductor must harness the talent of the various musicians and coordinate their actions so that they play the correct notes at the right time to produce music.

The brain acts in a similar fashion. It must simultaneously control the operation of all the body's organs and handle the input of electronic transmissions from the nerves that carry stimuli relating to the senses. The brain must constantly monitor all the activities of the body.

The Specialists

When it comes to the people who study and treat the brain, you can't tell the players without a program. Here's a brief primer:

> *Molecular biologists* do not restrict their work to the brain, but their research is extremely important to understanding its structure and operation. They look at what takes place in the nerves and cells at the molecular level.

Neurophysiologists try to understand the interactions of the various components of the nervous system.

Psychologists study the behavior and pathology of the brain as a whole and treat abnormal behavior without the use of drugs or surgery.

Psychiatrists are medical doctors who study behavior related to brain pathology and can prescribe drugs as well as other forms of therapy.

Neurologists are medical doctors who diagnose and treat disorders of the brain, spinal cord, and peripheral nerves nonsurgically.

Neurosurgeons are medical doctors who diagnose and surgically treat disorders of the brain, spinal cord, peripheral nerves, and their supporting structures such as the skull, the spine, and the blood vessels supplying these areas.

A more recent addition to the army of phy-sicians engaged in the treatment of brain disorders is the *neuroradiologist*, a medical doctor interested in the diagnosis and, in many cases, the treatment of abnormalities of the blood vessels of the brain. In fact, you can tack "*neuro*" in front of a lot of specialties and get a subspecialty related to the nervous system.

Daily Revelations

Nearly every week, the newspaper carries a story about some new development in brain research or the treatment of brain-related disorders. While this book was being written, for example, the local papers had the following headlines:

"Music Stimulates Brain Like Sex, Researchers Find." This report said people may have strong emotional responses to music because the music stimulates the same parts of the brain as food and sex do.

"Study Finds Brain Reacts to Sex-Specific Chemicals." Certain chemicals similar to the male and female sex hormones reportedly triggered distinctive brain activity when sniffed by the opposite gender.

"Spirituality on the Brain." This article discussed new research that shows what a part of the brain may be doing when some people are experiencing a "religious" feeling.

"Are Teens Just Wired That Way?" Noted research suggested the brain changes during the teen years.

"Study: Brain Growth Does Not Stop in Adolescence." This hopeful report described research indicating that a key element of brain development continues until nearly age 50.

"Is Soy Bad for Brains?" This report was on a study showing that soy hastens mental deterioration.

Almost everything you read in the newspapers, and particularly the tabloids, is either old news in the medical community or unproven. As you can see from this handful of examples, stories about the brain range from the wildly optimistic to the depressing to the downright dangerous. You should always check with a physician concerning anything that sounds like an extraordinary development or medical cure.

The Great Unknown

The brain is complex, and over the centuries that complexity has been translated to knowledge about the structure and function of the brain. Still, we know remarkably little about the brain. In the next few chapters, we're going to share with you a good deal of what is known.

Gray Matter

"The brain is a world consisting of a number of unexplored continents and great stretches of unknown territory."

—Santiago Ramon y Cajal

Some of your questions may be left unanswered because scientists and physicians have been searching for centuries to find answers to those same questions. No one truly understands the miraculous functioning of the brain, but you're going to be a whole lot closer in about 300 pages or so.

The Least You Need to Know

- Your brain is small and squishy; it is comprised mostly of water and uses one fifth of your body's blood and 20 percent of its energy.
- Darwin discovered that the fittest species survive and that related organisms descended from common ancestors.
- Most human brain development occurs in the womb and in the first two years of life. By age five, the brain is roughly the size and weight it will be for life.
- The brain is often compared to a computer, but it is far more complex than any machine.
- Don't believe everything you read in the paper. Medical news is often old or unproven. There is still more we don't know about the human brain than we do know.

Chapter 2

From Stone Age to Sculptors

In This Chapter

- Stone Age surgeons are successful
- Egyptians describe and discard brains
- Greeks philosophize about the head and the heart
- Renaissance men discover anatomy
- Barbers cut deeper

Humans have always been fascinated with brain anatomy and function. Even ancient cave petroglyphs show a caveman striking another over the head with a large club. The caveman must have been aware of some type of brain function because he realized this act would render the individual unconscious or dead.

Aha! The first neurophysiological experiment!

Prehistoric Brain Surgery

Given that the brain is perhaps the most complex machine in the universe, you may be surprised to learn that brain surgery is one of the oldest medical treatments. Apparently, humans didn't start out trying to heal each other with herbs or magical potions or leeches. Instead, they got right to work drilling skulls.

Not only is there evidence of brain surgery dating back to about 7000 B.C.E. (the late Stone Age), but also these were *successful* operations!

Gray Matter

The first known mention of brain physiology comes from Sumerian records that report that poppy plants (from which heroin is derived) cause euphoria when ingested. These records date back to 4000 B.C.E.

Words of Wisdom

Trephining was an ancient practice of cutting holes in the skull, which may have been intended to release from the brain or mind the evil spirits and demons believed to cause mental and physical illnesses.

IQ Points

In Europe, the piece of bone cut from the skull was sometimes worn around the neck as an amulet to ward off evil.

Archaeologists have discovered preserved skulls with holes in them and some surgical instruments of these medical pioneers in France, but brain surgery was not a strictly European phenomenon; it was practiced throughout the world. Evidence of brain surgery in Africa dates to 3000 B.C.E., and as early as 2000 B.C.E., Peruvians predating the Incas practiced the same form of surgery, which is known as *"trephining."*

This method of brain surgery was used to cure people of mental illness, headaches, epilepsy, and other illnesses as well as for spiritual and magical reasons. The practice continued to be used for centuries and was still in vogue in Europe as late as the sixteenth century.

Sometimes multiple holes were drilled in a single skull, but they were generally made with surprising neatness given the crude nature of the tools, which included wooden sticks with flint tips, bronze knives, and scalpels of copper or volcanic glass. Today, holes are still drilled in people's skulls to relieve pressure on the brain, to repair fractures, to do biopsies, and to gain access to a tumor or blood vessel abnormality, but precision, electric or airpowered drills and saws are used for modern brain surgery.

How do we know that these patients with holes in their heads survived, let alone were cured? We know many patients survived because some of the trephined skulls show signs of healing. One poor fellow, for example, was found to have five holes, but only the last scar showed any signs of infection.

Mummies Weren't Too Bright

Though early humans apparently saw the need and benefit to opening up the heads of sick people, they did not give much thought to the importance of the brain. The ancient Egyptians, for example, gave it no respect at all.

The Egyptians believed in immortality; they thought the soul would return to the body after death to continue its earthly life. To prepare for this rebirth, the Egyptians would preserve dead bodies and the deceased's possessions. Starting about 3500 B.C.E., they began to embalm their dead, removing the major organs and placing them in jars before wrapping the body in linen bandages. Two organs were not treated this way: The heart was left in the body and often protected by an amulet because it was believed to be the center of a person's being and intelligence; the brain, which was thought to have no value, was unceremoniously scooped out through the nose and thrown away.

Ironically, despite their lack of respect for the brain, the Egyptians were the first ones to use the word. The brain is mentioned seven times throughout the 3,000 year-old Egyptian papyrus that contains the first written record about the nervous system.

This papyrus describes 48 surgical cases. Among them are 27 instances of head injuries and one case of injury to the spine. A number of these cases are important to neuroscience because they discuss the brain, meninges (coverings of the brain), spinal cord, and cerebrospinal fluid for the first time in recorded history.

Case 6, for example, refers to a head wound with a skull fracture and opening of the meninges. It describes the convolutions of the brain as being "like those corrugations which form molten copper." The author concludes that this wound is "an ailment not to be treated."

Another untreatable wound was described in Case 8 in which a fracture of the skull led to paralysis of the arm and leg on the same side of the body as the injury. The first documentation of aphasia, in which a fracture of the temporal bone in the skull left the patient in Case 22 unable to speak, predated the famous work on aphasia by Paul Broca (1861) by thousands of years.

Gray Matter

On January 20, 1862, in the city of Luxor, Edwin Smith made an important historical discovery when he bought an ancient papyrus from a dealer named Mustapha Aga. The papyrus turned out to be the first written mention of the nervous system. After Smith died, his daughter, Leonore Smith, gave the papyrus to the New York Historical Society. In 1930, Henry Breasted published the English translation of the papyrus.

IQ Points

The first known physician was the Egyptian Imhotep (circa 2686–2613 B.C.E.). After his death, he was worshipped as a god for his healing powers. He is not to be confused with the evil priest with the same name who returns from the dead in the film *The Mummy*.

The Great Greeks

The Greek physician Hippocrates (460–370 B.C.E.) is considered the father of modern medical ethics. The son of a physician, Hippocrates was born on the Aegean Island of Cos. Roughly 70 texts are considered part of his collection, but no one knows which, if any, were actually written by Hippocrates.

Hippocrates believed the brain, or part of it, was a gland that cooled the blood and secreted the mucus that flows from the nose. His works also demonstrate that the author had knowledge of head injuries by accurately describing spasms and seizures and classifying fractures and contusions. Trephining is described and recommended for fractures of the skull. The descriptions of cases are also remarkable for their time; nothing comparable appears for nearly 2000 years.

Hippocrates left many texts on brain surgery, but he never operated on anyone with a skull fracture. Though he clearly saw a role for surgery, Hippocrates believed in the healing power of nature. He is probably best known for the oath attributed to him that offers physicians to this day an ethical guideline for the practice of medicine. It begins, "First, do no harm." Hippocrates was also important to the history of medicine because he helped move the field from a belief in supernatural reasons for illness to a greater acceptance of physical causes.

> **CAUTION**
>
> ## Code Blue
>
> The Babylonian King Hammurabi wrote the first medical rules about 2000 B.C.E. These rules established a reward system that could put an end to medical malpractice. For example, a physician received 10 shekels of silver for curing a nobleman's wound or eye abscess, but if he caused the nobleman to die or to lose an eye, the physicians' hands were cut off. Talk about a fierce penalty for malpractice!

Heart vs. Head

The Greek philosopher Alcmaeon of Croton (circa 450 B.C.E.) operated on eyes and discovered passages from the sense organs to the brain. He was one of the first people to believe the brain was the seat of thought and feeling.

Another Greek physician's son, Aristotle (384–322 B.C.E.), was better known for his works of philosophy, but he also contributed to medical science. He was a great observer who dissected animals and described many of the body's organs and produced some of the earliest anatomical drawings. Though his teacher, Plato, agreed with Alcmaeon's view, Aristotle maintained that both thought and feeling were based in the heart and that the brain's role was to cool the heart to prevent it from overheating. He also believed that the cooling agent was phlegm.

> ### IQ Points
>
> This book focuses mainly on the medical developments in the Western world. Advances in medicine were made in the East as well, but very few of these advances were related to the brain. Nei Ching described some body parts more than 2,000 years ago, but the Chinese generally placed less emphasis on the structure of the body than on the spirit. Like the philosophers and physicians of the West, the Chinese believed in unseen forces that influenced behavior and health. They focused on the energy or *chi* that flowed through body channels they called meridians. This energy balanced the female properties (the yin) and the male ones (the yang).

Herophilus (circa 335–280 B.C.E.) believed that the cerebral ventricles were the source of intelligence. Herophilus also recognized the brain as the center of the nervous system, distinguished the motor from the sensory nerves, and accurately described several of the internal organs.

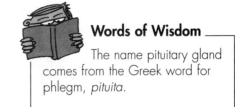

Words of Wisdom ____

The name pituitary gland comes from the Greek word for phlegm, *pituita*.

Another Greek, Erasistratus of Chios (304–250? B.C.E.), found that organs were associated with systems of veins, arteries, and nerves, which divided into smaller and smaller branches. He believed that air was taken into the lungs and passed to the heart, where it was changed to *pneuma*—the vital spirit—that was sent to other parts of the body via arteries. It reached the brain through ventricles and then changed into a new form of pneuma, which he called the animal spirit. This spirit passed through the body in what he believed to be hollow nerves.

Erasistratus also made pioneering observations about the brain, noting its convolutions and distinguishing between parts of the brain such as the cerebrum and cerebellum. He described the cerebral ventricles within the brain and the meninges that cover it. Erasistratus also postulated that the reason the brains of humans were more convoluted than those of animals was related to the higher intelligence of people.

IQ Points

The medical symbol showing a staff entwined by a snake, which is known as a caduceus, originated during World War II when medics used it as a symbol for a flag of truce. In ancient Greece the symbol was associated with the god-healer Asclepius. The caduceus was the staff carried by Hermes, the messenger of the gods, as a symbol of peace. It had two snakes around it while that of Asclepius had one.

A Sense of Humors

The Greeks also came to believe that all things were made up of four basic elements: air, water, fire, and earth. These elements were mirrored in the body by four liquids they referred to as humors: blood, phlegm, yellow bile, and black bile. According to the Greeks, temperament and illness were related to the balance of humors, and this remained conventional wisdom until the seventeenth century.

Yet another Greek physician, Galen (129–199 C.E.), had perhaps the greatest influence of any of the early medical researchers. He was a surgeon in a school for gladiators and later physician for three Roman emperors.

In his research, Galen dissected animals and became especially famous for his descriptions of other parts of the body, especially bones and muscles, rather than his work on the brain. He did, however, conduct experiments that helped to provide an understanding of the nervous system, such as cutting nerves to study their function and finding that, for example, severing the laryngeal nerve affected speech. Another important contribution was to disprove the idea of Erasistratus that air carried from the lungs to the heart is converted into a vital spirit distributed by the arteries. Galen showed that arteries carried blood not air.

Galen influences 45 generations.

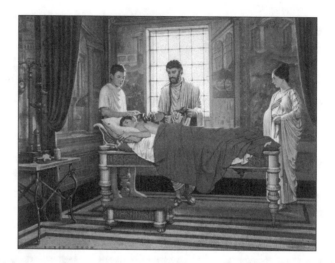

Still, much of Galen's work turned out to be wrong, in part because he did not have human subjects and relied instead primarily on dissecting apes. Thus, he detailed a network of tiny vessels at the base of the brain that he called *rete mirabile* ("wonderful network"), which exists in many animals but not in humans. Many of Galen's errors were perpetuated for nearly 1,500 years.

Science Takes a Holiday

After Galen's death, anatomical and physiological research came to an end and the Dark Ages of medicine began. The void lasted for about 1,300 years.

Interestingly, the mighty Romans, who brought so many advances to the world, made very little contribution to medicine. The Romans left their mark in the areas of hygiene and health services, notably the development of the first hospitals, but none of their physicians left a lasting legacy.

A major reason for the decline in scientific research was the growth of Christianity and influence of the church, which banned the study of *anatomy*. Rational medicine and faith healing had often existed together, but as the church's role became more central to political and social life in Western Europe, it became more and more difficult for scientists to explore areas that were seen as the province of God. Worse yet, as some of the Renaissance scientists would learn, was to suggest that humans might be guided by some power that was short of the divine or that varied from the perspective of the theologians.

The general Christian belief was that disease was a divine punishment for committing sins and should be treated by prayer, superstition, healing saints, or other religious means. The real concern was about the fate of the soul after death rather than the preservation of life. The dogma of the Middle Ages was that the body was a temporary home for the soul and that the physical matter inside the body wasn't important.

The tension between science and religion continues to this day in debates over issues such as brain stem cell research. Just as religious concerns have impeded such modern medical study, so too were they a negative influence in earlier times.

IQ Points
The Romans are credited with the development of one surgical procedure that is still used today: the caesarean section. The operation was originally performed when the first king of Rome ordered the opening of a dying woman's body to extract a child. The name comes from the story that Julius Caesar was born this way.

Words of Wisdom

The word **anatomy** comes from the Greek *anatom*, which means dissection.

A more congenial relationship existed between philosophers and physicians. Many scientists, such as Aristotle, were also philosophers. This association was common for centuries. These two fields also continue to overlap today, particularly when it comes to the study of the mind.

From Darkness to Light

Although the church generally impeded medical research, various religious organizations did establish hospitals and other medical institutions to care for the ill in Western Europe. In addition, religious orders, especially the Benedictine monks, who were among the few people who could read and write at that time, took great care to collect, study, preserve, and copy ancient medical texts. The greatest of these collections was probably at Monte Cassino near Salerno, Italy. In the ninth and tenth centuries, this area became the center of medical education and care in Europe and home to the first medical school.

Meanwhile, further east, the Romans were supplanted by the Muslims from Arabia who created a vast empire from Persia across North Africa and ultimately into Spain. They, too, preserved the work of the Greeks and translated it to Arabic so it could be disseminated throughout their empire. Later, it was retranslated back into Latin, and the medical knowledge was returned to Europe. In the meantime, several Islamic physicians made important contributions to the study of medicine. These included important philosophers, al-Razi (852–932), perhaps the greatest Islamic surgeon; Abulcasis (930–1013); and Avicenna (980–1037). The Jewish thinker Maimonides (1135–1204) was an important contributor as well. None of these people, however, made a significant discovery related to the brain.

The *Renaissance* began in the fourteenth century and lasted for roughly three centuries. During this period, the people of Europe began to question many of their long-standing beliefs. Christianity remained a dominant force in politics and culture, but people gradually shifted away from focusing on life after death to take a greater interest in the world around them.

Though scholars now question whether this period marked as dramatic a change in human thought and behavior as once believed, the Europeans were confident that they were creating a whole new culture. And although the Middle Ages may not have been as dark and backward as people once thought, there is no question that the Renaissance was a period of intellectual ferment that laid the groundwork for modern science.

Words of Wisdom

The word **renaissance** means "rebirth." The Europeans believed they were rediscovering Greek and Roman culture after centuries of intellectual and cultural decline and returning civilization to greatness.

The great artists of the Renaissance, such as Michelangelo, Raphael, and Leonardo da Vinci closely studied the human form because of their desire to represent it accurately. The prototype of the "Renaissance man," Leonardo da Vinci (1452–1519), was one of the first to question the conclusions of Galen. Philosophically, da Vinci believed in very careful observation because he was convinced that outward behavior revealed inner thoughts. In fact, his paintings are often described as being "psychological portraits" of their subjects.

One of the keys to da Vinci's contribution to science was his conviction that the world could be understood by studying it objectively. This was a radical concept for its time because the prevailing view was that many things were spiritual in nature and anything that defied rational explanation could be attributed to the supernatural.

Da Vinci challenged the views that Galen had propounded—and that had been accepted virtually unchallenged for centuries— namely that the balance of the four bodily humors (blood, black bile, yellow bile, and phlegm) were responsible for the health of the body. Since this was all a physician really needed to know, there was no apparent need for a detailed study of the body's structure.

As da Vinci turned away from his artistic inclination, he became intensely interested in the functions of the body. He performed research on animals, dissected human corpses, and conducted physiology experiments. From his observations he made some of the most impressive medical drawings and insights into anatomy. da Vinci was particularly interested in the heart, lungs, and brain, which he saw as the "motor" of the senses and of life. It is difficult to appreciate the difference between what came before and da Vinci's diagrams without seeing them, but imagine a relatively accurate drawing of a person's body with something that looks like a street map of maze from a puzzle book drawn over it. That was the old school of anatomical drawing. By contrast, one of da Vinci's would closely resemble a diagram in any modern medical text.

Beyond this general study, da Vinci also made very specific inquiries into the nervous system. He drew pictures of the peripheral nerves, for example, and recognized relationship between some nerves and the senses. Da Vinci held to the notion that the soul was in the brain, but also saw the brain as a control center for the body. By injecting wax into a brain, da Vinci discovered the ventricles, cavities deep in the brain where he believed, incorrectly, the nerves led. Da Vinci also described reflexes and apparently understood that they involved the spinal cord and not the brain.

Despite all these insights, da Vinci had no influence on his contemporaries, however, because his work was not discovered until centuries later. For example, Leonardo made casts of the ventricles of the brain to get an accurate idea of their shape. Apparently he was going to publish a book on anatomy, but he abandoned the project when his co-author died.

One of the most important Renaissance figures in the field of medicine was the Flemish physician Andreas Vesalius (1514–64), who revolutionized the study of biology and the practice of medicine by his observations and descriptions of human anatomy. Unlike most of his predecessors, he was not restricted to dissecting animals, and his study of human corpses allowed him to write and illustrate the first comprehensive human anatomy textbook. In addition, he demonstrated that many of the accepted views first put forward by Galen 1,400 years earlier were incorrect.

Vesalius's work on the brain was less significant, though he did include a number of excellent diagrams in his book and described the ventricles, medulla, and cerebellum. His work, like that of Galen, still was limited by the research methods and technology of his time, so he still made some mistakes.

> **Gray Matter**
>
> Vesalius's seven-volume anatomy textbook, *De Humani Corporis Fabrica* (On the Structure of the Human Body) was published in 1543, the same year another revolutionary work of science appeared, Copernicus's *De revolutionibus orbium coelestium libri vi* (Six Books Concerning the Revolutions of the Heavenly Orbs), which said, among other things, that the earth revolved around the sun. Unfortunately for Copernicus, he died a year before his book appeared.

Other lesser-known physicians and scientists made contributions to the study of the brain. For example, in 1518, Laurentius Phryesen published a book that included six pictures showing dissections of the brain at different stages. Jacobo Berengario da Carpi described several important structures deep inside the brain: the ventricles, choroid plexus, and the pineal gland. Bartolommeo Eustachi (1520–74) made illustrations of the base of the brain and the sympathetic nervous system.

Even in the Renaissance, however, the brain was still not viewed as related to thinking, emotion, and intelligence. At the time of Shakespeare, in sixteenth-century England, the liver was believed to be the site of these functions.

A Form of Barbarism

Though today surgeons are often thought of as the elite among the medical community, their status was considerably lower centuries ago. Strangely enough, surgeons often were also barbers who traveled from town to town cutting hair, pulling teeth, stitching wounds, and bloodletting (the practice of draining blood from the body, which was thought to cure illness). The red-and-white striped pole outside old-fashioned barber shops is a vestige of the connection between barbers and surgeons. The red stripes represent blood, and the white stripes symbolize bandages.

In an effort to improve their training, regulation, and status—not to mention their image!— surgeons began to establish colleges, the first of which was the Royal College of Physicians, founded in London in 1518.

As the modern age of science nears, it is worth noting that as ridiculous as some of the ancient theories may sound today, they made sense in the context of their times. Future scientists may look back at our work with the same amusement that we view the old beliefs in humors and the role of the liver in intelligence.

The Least You Need to Know

- Brain surgery dates back thousands of years when people began drilling holes in the heads of sick people.
- The Egyptians were the first to mention the brain, but didn't think it was an important part of the body.
- The Greeks studied the body and made many discoveries, but most of them remained convinced that the heart was the seat of thought and feeling.
- The Renaissance brought a renewal of interest in the body and a host of new advances in anatomy, particularly with regard to describing the brain.

3

Knowledge of the Brain Evolves

In This Chapter

- ◆ Mind versus body
- ◆ Quackery
- ◆ Pavlov's dogs
- ◆ Anesthesia and antiseptics
- ◆ Darwin the evolutionary

The Renaissance may get a lot of credit for its artistic influence on the world, but the period was not especially important to the advancement of medicine. The medical sciences finally began to pick up steam in the seventeenth, eighteenth, and nineteenth centuries. Before the breakthroughs in describing and understanding the brain were achieved, however, a more philosophical question intruded.

I Think Therefore ...

Through the centuries, the question of where the soul resides has occupied the thoughts of many philosophers. Some believed the heart, others the liver, and a few thought it might be in the brain. Fewer still believed the brain was important to thinking or intelligence. Then a French philosopher and mathematician named René Descartes (1596–1650) came along and introduced a slightly different, but no less vexing, issue.

According to Descartes, a human is composed of two basic elements: body and mind. The body is like a machine and is tangible and measurable, while the mind is more abstract, invisible yet capable of thought. Given the limited understanding of the brain's function, and almost total ignorance of its role in processing information, it is not surprising that this duality struck a positive cord.

If you accept that the mind and body are distinct, the next question is how they are related. According to Descartes, the pineal gland, located deep inside the brain, was the key terminal where the mind exerted control over the body. He chose the pineal gland because he believed (incorrectly) that it was uniquely human and because he thought it was the only part of the brain that was not duplicated.

Vision and the mechanism for response to external stimuli as seen by René Descartes.

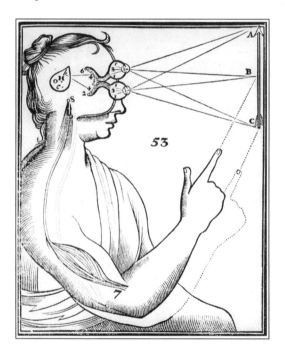

If you want to learn more about the intricacies of what is called Cartesian dualism, you'll have to pick up *The Complete Idiot's Guide to Philosophy.* For our purposes, what's important is that the Frenchman got people thinking about the relationship between the brain and mental processes.

One problem with Descartes's philosophy, from the perspective of medicine, was that it discouraged investigation of the mind because of the alleged inability to scientifically study it. It took about 300 more years before scientists began to find evidence that human qualities associated with the mind, such as personality, may be determined by biochemical changes.

A Place in the Heart

While Descartes was thinking about the nature of the mind and body, others were conducting physical experiments. The seventeenth century ushered in a new era of scientific research during which Galileo, Kepler, and Brahe made their great innovations in astronomy, William Gilbert demonstrated the properties of magnets, and Robert Boyle (1627–1691) helped discredit *alchemy* and introduce the field of chemistry. Boyle's experiments helped produce a greater understanding of the composition of the living world and undermine the long-standing beliefs in humors and spirits.

> **CAUTION**
>
> **Code Blue** _____
>
> Although the so-called mind-body problem Descartes raised still provokes philosophers, his theory about the role of the pineal gland has been thoroughly discredited. The gland is involved in sleep and wakefulness and has nothing to do with the mind.

> **Words of Wisdom** _____
>
> **Alchemy** was a practice in which practitioners claimed the ability to mysteriously transform ordinary materials into something special. Some alchemists believed they could change base metal into gold, cure diseases, and indefinitely prolong life.

Perhaps the most important development in medicine in the seventeenth century was the discovery by Britain's William Harvey (1578–1657) of how the blood circulates through the body. This discovery has nothing directly to do with the understanding of the brain, but it indirectly led to the investigation of aspects of the body that did result in advances in knowledge of the brain. In 1658, for example, Johann Wepfer theorized that a broken blood vessel in the brain may cause apoplexy (stroke). During the remainder of the seventeenth century, the only major advance in the study of the brain was the description and naming of the fourth and eleventh cranial nerves.

In 1664, the English anatomist Thomas Willis (1621–1675) introduced the term *cerebel* for what we now call the cerebellum and correctly hypothesized that it was responsible for unconscious movements. He also wrote the first book on the anatomy and physiology of the brain, which included a description of the vascular interconnections at the base of the brain that came to be known as "the circle of Willis." He also studied the functions of individual parts of the brain and suggested that different areas controlled specific functions, distinguishing, for example, between the parts involved in thought and motor function. Willis has been called the father of neurology, a term he also introduced, along with "hemisphere," "lobe," and "peduncle." Willis wrote a second influential book, *De Anima Brutorum*, which described psychological and physiological disorders such as vertigo, apoplexy, paralysis, and delirium. In addition to these brilliant insights, however, Willis also erroneously believed in "animal spirits" that were distinct from the brain.

Gray Matter

In 1691, Robert Boyle mentioned the case of a knight who suffered a depressed fracture of the skull and lost sensation and muscular movement in an arm and a leg. A surgeon removed a sharp piece of bone from the knight's brain, and the paralysis disappeared in a few hours. This result suggested that an area on the surface of the brain might control motor function.

Out of Sight

The old saying, "out of sight, out of mind," can be applied to the study of the body. For centuries, scientists had to rely on their unaided eyes to observe the structures of the body. Just as Galileo's invention of the telescope helped revolutionize astronomy by making it possible to see parts of the sky never viewed or imagined before, so, too, did the microscope revolutionize medicine by allowing researchers to discover the tiniest building blocks of the human body.

The microscope was invented at the end of the sixteenth century and was used by early microscopists such as Antoni van Leeuwenhoek (1632–1723) to see blood cells, muscle fibers, and other minute components of the body for the first time. Some of the organisms these scientists saw would later be identified as bacteria and recognized as a cause of infectious diseases.

Italian anatomist Marcello Malpighi (1628–1694) was another pioneer in the use of the microscope and was the first to describe the cerebral area of the brain. A number of anatomical structures elsewhere in the body now bear his name.

The quest to explain thought remained a preoccupation throughout the eighteenth century. Julien Lamettrie (1709–1751) was influential in suggesting that humans are like machines. He believed that a machine can continue to work, albeit with some problems, even if it is missing certain components. Similarly, a human could function even if some parts were missing. He argued that the brain is essentially a blank slate and that it secretes thoughts the way the liver secretes bile. He compared the thought process to a violin string that vibrates to produce sound. The cerebral fibers, he said, are struck by sound waves that stimulate a repetition of what strikes them.

What Do You See?

One long-standing mystery is how we are able to see and translate what comes into our eyes into a recognizable image. In the first half of the eighteenth century, a lot of thought and experimentation was devoted to this question.

William Molyneux (1656–1698) and an Irish Anglican Bishop George Berkeley (1685–1753), for example, were interested in the question of what would happen to people who had been born blind but later regained their sight. While blind, the people had learned to recognize objects largely by touching them. Once they could see, however, they learned how to identify shapes and sizes and distances. Would these people be able to identify an object that was familiar to them when they were blind by sight only, without touching it? If they could, then their recognition of the object had to be related to their experience of the object, not just its appearance.

Philosophy and brain function came together again after Descartes's death with the introduction of the "Molyneux Problem." Irishman William Molyneux (1656–1698) wrote a letter to his friend, the English philosopher John Locke (1632–1704) in 1693, asking a theoretical question: If a person blind since birth, suddenly regained sight, could they distinguish objects by sight alone? For example, if the person knew the difference between a square and a sphere by touch alone when they were blind, would they recognize the difference when they saw them for the first time? Molyneux believed they could not because seeing is learned and it is the soul, not the eye that sees.

The Irish Anglican Bishop George Berkeley (1685–1753) agreed that vision was learned. Berkeley believed there was no connection between our individual senses. It is only through experience that we learn to associate different types of stimuli—such as color, touch, light, smell, and sound—with a particular object.

In 1749, David Hartley (1705–1757) suggested that when sense organs were stimulated, the nerves would vibrate and send messages to the brain, which translated them into ideas. This suggestion begged the question of how this translation was carried out. Hartley argued that the vibrations hung around inside the brain so that even after the initial cause of the sensation was gone, it would be remembered.

If sensation was the cause of thought, how could you account for different people interpreting the same sensations in different ways? Philosophers offered a variety of ideas, from the way an individual's brain connected ideas to the quality and quantity of someone's education.

Gray Matter

George Berkeley was a proponent of the philosophy of idealism; this philosophy asserts that nothing, including material objects, exists apart from perception. All objects, he argued, are collections of ideas and sensations.

Quack, Quack

Throughout medical history scientists and pseudo-scientists have proposed treatments for disease that range from the absurd to the ridiculous. Some treatments, though unconventional, attracted loyal followings and retain a measure of credibility even today. The ancient Chinese practice of *acupuncture* and the eighteenth-century systems of *homeopathy, aromatherapy,* and other forms of alternative medicine are among the different treatments adopted over the years.

Words of Wisdom

Acupuncture is an ancient system of healing developed in the Far East that involves the insertion of fine needles into specific locations of the body. **Homeopathy** was developed by a German doctor named Samuel Hahnemann (1755–1843), who believed that chemicals that caused diseases could act as cures if used in extremely small doses. **Aromatherapy** uses oils from plants to treat a variety of disorders. The fragrant substances are sometimes massaged into the skin and other times inhaled.

In some cases, however, out-and-out quackery developed. There was nothing special about the eighteenth century that should have stimulated a burst of quackery, but two of the most unusual ideas relating to the brain occurred at this time.

A Galling Theory

Franz Gall (1758–1828) studied the skulls of dead people and tried to draw relations between their appearance and the individual's personality when they were alive. He decided that the size of the bumps on the skull corresponded to more than two dozen character traits, such as friendship, sense of humor, and goodness. Gall's method was to put a kind of hat on subjects' heads that had movable pins that would create perforations in a piece of paper. Gall would then use this map to determine the person's character. The theory became known as phrenology ("the study of the mind").

For several decades phrenologists read the heads of patients and published their results in books and periodicals. In the United States, the foremost phrenologists were the Fowler brothers, who began "reading heads" in New York in the 1840s.

A HINT TO PHRENOLOGISTS; or, "September 20, 1878."

OPERATOR IN STOCKS (who desires a chart of his head, interrogatively). "I say, Professor, it cuts to me, from the number of skulls I see, that you will have rather a lively time in this Establishment on Resurrection-Day?"

PROFESSOR (who can find no traces of the bumps of veneration). "No; we shouldn't anticipate any trouble, except, perhaps, from those Wall Street Skulls on the second shelf: they might appropriate other people's bones, and create a temporary panic."

Phrenologist examines the patient's head for meaning.

Look into My Eyes

Instead of the bumps on the head, Franz Anton Mesmer (1734–1815) believed that the planets were the greatest influence on behavior and that magnets could have a similar behavioral affect. After awhile, he decided he could dispense with the magnets and simply use his hand to affect the body's "animal magnetism." Apparently he put on quite a show, using his power to cause patients to appear to be asleep but still respond to his commands and be capable of movement. This was the first known use of *hypnosis*. After the patients woke up, they said, or Mesmer claimed, that they were cured of whatever ailed them.

Mesmer convinced enough people that he could change the flow of magnetic fluid in patients, and hence their behavior, that he became rich and famous. His fame also prompted a more serious investigation into his claims. In 1784, the French Academy of Sciences concluded that "animal magnetism" did not exist. Mesmer subsequently lost both his wealth and his reputation, but his name lives on in the word mesmerize.

Though Mesmer's animal magnetism theory was debunked, others pursued the idea that a person

Words of Wisdom

The term **hypnosis** was coined by James Braid around the middle of the nineteenth century. It is also referred to as mesmerism after Franz Anton Mesmer and describes a condition that resembles sleep but is induced by another person.

could be put in a hypnotic state. In 1820, Alexandre Jacques Francois Bertrand (1795–1831) began to study hypnosis, and gradually it became accepted as a means for treating certain mental illnesses, alleviating pain, and countering anxiety. More controversial is the notion that under hypnosis people can remember things they've forgotten or unpleasant memories they subconsciously buried. Physicians and dentists sometimes use hypnosis, but it is a more common technique among psychologists and psychiatrists, who employ it to treat a variety of mental problems.

Code Blue

> Not everyone can be hypnotized. Some people are more susceptible than others, though no one is sure what determines a good subject. Willingness or belief do not appear to matter. Also, despite the impression given in movies, hypnotists cannot program people to do their bidding or to commit crimes. Though stage hypnotists might get people to do silly things like hop on one foot or bark like a dog, they can't make people act contrary to their morals.

You're Electric

The latter part of the eighteenth century also had its share of important discoveries. In 1773, John Fathergill described trigeminal neuralgia, one of the first delineations of a neurological disorder. The trigeminal nerve is the fifth of the twelve cranial nerves and has a direct attachment to the brain, so trigeminal neuralgia is a condition characterized by a stabbing, shock-like facial pain caused by a lesion or compression of the trigeminal nerve.

Luigi Galvani (1739–1798) opened a whole new area of research by using electricity to stimulate the nerves of frogs. He showed that what he called "animal electricity" was a force within the body, including within brain cells. Galvani's was one of the many accidental discoveries that have occurred in history. His metal equipment created an electrical cell that touched the bodies of animals he'd recently dissected. When he saw that it caused their muscles to twitch, he made the connection between outside electrical stimulation and body movements. According to another story, Galvani laid a frog out on a metal plate and noticed its legs twitched at the sound of thunder and lightning.

One of the consequences of Galvani's discovery was to make the use of a variety of electrical instruments a fad in medical treatments. Most of these practices were shown to be useless or dangerous and were abandoned.

It took almost 60 more years before another scientist could perform more exact calculations of the electric currents. Emil Du Bois-Reymond (1818–1896) discovered that the electrical pulses from neurons traveled at NASCAR-like speeds of roughly 200 miles per hour.

The discovery that electricity had an influence on the body also prompted efforts by scientists and doctors to use it to try to bring the dead to life. The goal was usually to help people who drowned and had just died. This technique was used unsuccessfully on poet Percy Shelley's first wife, who drowned in London in 1816.

After Galvani's discovery, Shelley's second wife, Mary, began to imagine that a corpse could be reanimated using electricity. In her classic story, Dr. Frankenstein assembles a variety of body parts and creates a new being that so horrifies him that he wants nothing to do with it. The monster then seeks revenge against his creator by murdering those he loves. Most people are more familiar, however, with the Hollywood version in which the doctor puts the brain of a criminal into the body of a corpse he has dug up, which causes the monster to be violent.

Gray Matter

Shelley's story of Frankenstein has been retold many times. For my money, the best is Mel Brooks' hilarious parody, *Young Frankenstein*. In one classic scene, after his new creation turns violent, Dr. Frankenstein (played by Gene Wilder) asks his assistant Igor (the wild-eyed Marty Feldman) where he got the brain that has just been put in the creation. Igor tells him the name on the jar was Abby Normal.

Oops! I Mean, Eureka!

Small but critical pieces of the brain puzzle were put together in the nineteenth century, which helped accelerate the advancements in this area of medical science. Jan Purkinje (1787–1869), for example, discovered that nerve cells had a different structure from other cells in the body. One part, the nucleus, was similar to other cells; however, the fibers that extended from the nerve cell nucleus were new and different. Later, these parts would be called axons and dendrites.

Crucial to the understanding of the relationship between the nerves and the brain was Camillo Golgi's (1843–1926) discovery in 1873 that the neurons in the brain transmitted information to the motor nerves and that information from sensory nerves is conveyed to the brain. He discovered neurons by accident when he knocked a piece of an owl's brain into a solution of silver nitrate. A few days later he examined the specimen under a microscope and found that only certain cells, which turned out to be neurons, reacted to the stain. Golgi concluded that the nervous system was comprised of billions of these neurons.

Golgi thought all the neurons were connected like a giant web. In 1889, however, Santiago Ramón y Cajal (1852–1934) of Spain discovered that they were separated by tiny gaps called synapses. Rather than simply flow along a network of uninterrupted nerves, the electrical impulse had to cross the synapse. Cajal was also the first to isolate the nerve cells, called Cajal's cells, that are located near the surface of the brain. The combined work earned both Golgi and Cajal the 1906 Nobel Prize.

A further refinement of the understanding of nerves was the "Law of Specific Nerve Energies," propounded by Johannes Müller (1801–1858). Basically, this complex law says that the sensations produced by individual nerves are specific to that nerve. For example, the sensation of light is produced only when the optic nerve is stimulated. If the sensory nerves in your nose are activated, they will not produce a sensation of light, only smell.

More Than Just the Shakes

Earlier in the nineteenth century, James Parkinson (1755–1824) wrote his "Essay on Shaking Palsy" that described the symptoms that would later be recognized as part of a disease that would bear his name. Over 100 years later, Frenchman Charles Foix (1882–1927) discovered that the lesion responsible for Parkinson's disease was in a part of the brain called the substantia nigra.

Another syndrome, Bell's palsy, was discovered in 1821 when Charles Bell (1774–1842) observed facial paralysis on the same side as a lesion of the facial nerve. Another doctor, Dr. John Down (1828–1896) wrote a classic study on a congenital abnormality characterized by unusual facial features, including a protruding tongue, long ears, unusually round head, and decreased mental abilities, which later became known as Down's Syndrome.

The man some consider the founder of neurology, Jean Martin Charcot (1825–1893), also studied the nervous system and recognized several important diseases, including multiple sclerosis. One of his students, incidentally, was Sigmund Freud.

Finding the Controls

Little research could be done on humans, so researchers were largely limited to dissecting to observe the structure of their brains and conducting experiments to determine how injuries to different parts of the animals' brains affected their behavior and movements. Although many early experiments used species like frogs that were easy to obtain and study, scientists increasingly used higher order animals, that is, those with anatomies more closely resembling humans such as dogs and monkeys.

These animal experiments helped scientists learn that discrete parts of the brain controlled various body functions. For example, Luigi Rolando (1773–1831) conducted experiments in 1809 involving the removal of the cerebral hemispheres and cerebellum from animals. He also used electric current to stimulate the cortex. Based on his observations, he correctly concluded that the cerebrum controlled voluntary body functions and the cerebellum controlled involuntary functions.

Thirteen years later, another researcher, Marie Jean Pierre Flourens (1794–1867), added to the understanding of the cerebellum by removing it from pigeons and dogs. Flourens became the first to demonstrate that an injury to the cerebellum affected coordination.

English physiologist Richard Caton recorded weak electrical currents in the brains of rabbits and monkeys in 1875. This finding became important when later researchers found that chemical reactions cause electric currents that propagate all nerve impulses and transfer of information in the brain.

Caton's research was also important because it indicated the brain is naturally active; that is, it never rests. Later researchers discovered ways to record this brain activity, which allowed them to show the brain is not only active, but the degree of activity changes during periods of sleep and relaxation.

Code Blue

Beware of the snore! Novelist Charles Dickens described what we now know as obstructive sleep apnea in 1836. You may not have heard of it, but sleep apnea is as common as adult diabetes and affects more than 12 million Americans. Obstructive sleep apnea is caused by a blockage of the airway, usually when the soft tissue in the rear of the throat collapses and closes during sleep. With each apnea event, the brain briefly arouses victims from sleep in order for them to resume breathing; consequently, sleep is extremely fitful and of poor quality. Untreated, sleep apnea can cause high blood pressure and other cardiovascular disease, memory problems, weight gain, impotency, and headaches. Sleep apnea can be diagnosed and treated.

One of the main reasons for the extensive use of animals for research was that governments restricted the use of human bodies for the study of anatomy. In Europe, for example, it was forbidden until the seventeenth century to dissect human bodies for scientific research. This restriction stimulated the ghoulish practice of body-snatching, which is sneaking into graveyards and digging up corpses. The practice became so widespread that the laws were changed to allow the use of unclaimed bodies in medical research.

Pavlov's Dogs

Russian physiologist Ivan Pavlov (1849–1936) conducted the most famous experiments on dogs. He was interested in the process of digestion and was studying the relationship between the salivation of dogs and the actions of the stomach. He discovered that the stomach did not begin the digestive process unless the animal salivated. He then made the imaginative leap that ultimately helped him win the Nobel Prize.

Pavlov decided to test whether he could create conditions that would cause a dog to drool. He did so by ringing a bell at the same time he fed the dogs. Typically, the dogs would drool only when they saw and ate their food; however, after a period of time he observed that the dogs would begin to salivate when the bell rang, even if they did not have any food. Once the dogs learned the bell was related to being fed, they could then be taught that the two were unrelated by ringing the bell and not feeding the dogs. After awhile, the dogs would stop salivating at the sound of the bell. Pavlov called this behavior a *conditioned reflex*.

Pavlov's main interest was in the behavior of the body, but his work was enormously influential in the field of psychology. A whole school of thought called behaviorism emerged as an outgrowth of Pavlov's experiments. The behaviorists believe that heredity is not the sole or even most important determinant of behavior; rather, they maintain that emotions and many other behaviors are learned and determined by external influences. Today, we commonly use terms such as positive and negative reinforcement to describe rewards and punishments designed to change behavior. These ideas are outgrowths of behaviorism.

Words of Wisdom

A **conditioned reflex** is a learned behavior and is distinguishable from an innate reflex, which is automatic, such as pulling your hand away from a flame.

Of course many psychologists have other ideas, and some will be discussed in this book. The important point is that Pavlov and others demonstrated that everything we do is not simply determined by internal processes and that what the brain tells our bodies to do can be affected by what happens outside our bodies.

We're All Scientists Here

French physiologist Claude Bernard (1813–1878) had done important research on digestion that influenced Pavlov, but his more lasting contribution was his emphasis on the objectivity of experimentation and the necessity of proving or disproving a hypothesis. This concept is essentially what we now call the scientific method. This method was important because it led to the belief that the results of experiments were based on factual information rather than subjective ideas.

In London, pioneering neurologist John Hughlings Jackson (1835–1911) studied epilepsy and found that the movements associated with the disease were related to a part of the cortex.

Even more influential were the observations of Pierre Paul Broca (1824–1880). He had a patient who became known as Tan because that was the only sound he could utter. After Tan's death, Broca examined his brain and discovered a damaged spot on the front left side of the cerebrum that coordinates the muscles in the voice box and neck that produce speech. This spectacular breakthrough was the first time a particular activity was linked with a region of the brain. This part of the brain was later named for Broca. A few years after Broca's discovery, Carl Wernicke found that damage to another part of the brain caused aphasia, the inability to understand speech. His patients could speak, but they would invent new words or put words together in an incoherent fashion.

On our side of the pond, war was an important, though unfortunate, avenue for learning more about different types of injuries and their impact on behavior and body functions. The preeminent American neurologist of the time, Silas Weir Mitchell (1829–1914), served in the Civil War and wrote an important book on gunshot wounds and the related injuries to nerves. He also studied cerebellar function. In his spare time, Mitchell wrote romance novels and poetry.

Gray Matter

Broca and Wernicke's findings undermined the credibility of phrenologists who had assigned control of language functions to an entirely different part of the head; they said language was controlled by an area in the lower part of the left eye socket. On the other hand, phrenology helped stimulate the idea that some localization of cerebral functions might exist.

One of the most important medical developments during this time was the discovery of anesthetic agents. Prior to this discovery, surgery was necessarily very limited. Surgical procedures, the most common being amputation, usually required the patients to endure excruciating pain. Usually, doctors and nurses had to hold down screaming patients. For most of human history, the only way doctors knew to minimize pain was to give patients large quantities of alcohol or to knock them unconscious with a punch or a blow. Sometimes the pain was so great that the patients fainted.

Starting in the 1840s, chloroform, ether, and nitrous oxide (called "laughing gas" because it can induce laughter or hysteria) were all used to help reduce the pain patients felt during medical procedures. In 1846, Thomas Morton (1819–1868) used ether to render a patient unconscious while he removed a growth from the patient's neck. Shortly thereafter, ether and other gases that had fewer side effects became widely used during surgery.

Finding the Bugs

Another key discovery of the nineteenth century was the germ theory of disease. Though French chemist Louis Pasteur is best known for it, the German physician Robert Koch also was responsible for the idea that certain diseases are caused by tiny organisms that can be seen only through a microscope. As the causes of infectious diseases became better understood, vaccines and other means of preventing diseases that had ravaged civilization for centuries were developed. Some of these infectious diseases, such as rabies, affected the brain, so their prevention was important to eliminating or reducing the causes of brain disorders. (Pasteur created a vaccine to prevent rabies in 1885.)

After Pasteur showed that living organisms in the air caused decay, English surgeon Joseph Lister hypothesized that the microbes were getting inside wounds and causing infections. At the time surgeons were using the same basic techniques they had been using since the Middle Ages; they commonly operated in their street clothes with dirty instruments in filthy hospital rooms. Consequently, even if a patient were lucky enough to survive an operation, he or she was likely to die of a subsequent infection.

If the microbes could be killed before they entered the body, Lister believed, the infections could be prevented. Lister learned that carbolic acid was being used to treat sewage in one part of England and that a parasite causing disease in cattle was subsequently eliminated. He then used carbolic acid to clean wounds and discovered that it was an effective *antiseptic*.

> ### Gray Matter
> The mouthwash Listerine was named after Lister, and the product was originally designed as a disinfectant for surgical procedures.

> ### Words of Wisdom
> **Antiseptics** prevent or arrest the growth of microorganisms. **Asepsis** is a method by which harmful organisms are killed so they are never present in the operating room. For example, all furniture, instruments, and linens in a surgery room are treated with antiseptics.

Lister's innovation had its own drawbacks, particularly for the doctors who tried to implement it. Rather than use a chemical mist, which was the initial approach, many doctors put their clothes and instruments into high-temperature ovens that would kill the germs. This kind of sterilization became increasingly common after the invention of the autoclave in the 1880s. This approach is known as *asepsis*.

Before the germ theory was accepted, even simple steps such as washing hands and keeping operating rooms clean were revolutionary ideas. It wasn't until the 1890s that surgeons began to wear masks and rubber gloves. Though Lister's idea initially met with resistance, the gradual acceptance of antiseptics reduced the number of deaths caused by infection. Surgery became much safer as physicians learned to take greater sanitary precautions.

An Engaging Case

In 1848 an extremely unusual and interesting case came to the attention of physicians. Phineas P. Gage was a 25-year-old railroad worker. One day a gunpowder explosion blew a metal rod that was more than 3 feet long and weighed more than 13 pounds through his cheek and up through the front of his brain and out of his skull. The man, who literally had a hole in his head, and a large part of his brain destroyed, miraculously survived the accident and suffered no physical disability (he lived for 12 more years).

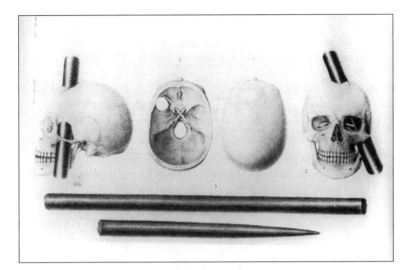

Gage's skull showing where rod passed through his head.

According to his physician, however, his personality changed radically after his injury so that he was "no longer Gage." For example, Gage had been a hard worker and a loyal friend who did not use profanity, but after his accident, he became lazy and quarrelsome and given to profane remarks.

Besides being a fascinating case study of someone who survived a terrible brain injury, the broader implication of what happened to Gage was that it suggested that the brain plays a role in personality. Gage's doctor specifically inferred that the damaged frontal lobe of his brain was responsible for rational decision-making and emotion.

"The Book That Shook the World"

The case of Phineas Gage helped foster a better understanding of the fully developed human brain. About the same time, thousands of miles away, British naturalist Charles Darwin was on a journey of discovery that changed our understanding of the origins of the brain.

On a voyage to the Galápagos Islands off the western coast of South America, naturalist Charles Darwin observed a number of unusual animal species that had not been seen before. He wondered how they had arrived at that remote island and developed their unusual characteristics. He also noticed animals that were more common, but had different traits across the different islands. Finches, for example, were similar sizes and colors, but they had a wide variety of beaks, which led Darwin to believe they all could be related to each other and may have had a common ancestor.

Darwin's conclusions regarding the way traits are passed on is one element of what came to be known as the theory of evolution. Darwin published his ideas in *On the Origin of Species.* The book sold out on the first day of publication in 1859.

In the context of understanding the brain, Darwin's theory led to the view that the brain itself has evolved from different species, starting with fish and reptiles and, later, from mammals. Different components of the brain also have grown and become more specialized to allow animals to adapt to their environments. Birds of prey, for example, have to have acute vision to see from high above the ground, so the parts of the brain related to vision are more highly developed. The cortex, the part of the brain most related to thought, is most highly developed in humans because of our advanced mental processes.

Real Brain Surgery

While Darwin, his followers, and his critics were debating where we all came from, others remained focused on the human brain as it had already evolved. In the mid-1870s, William Macewen (1848–1924), a Scot who had studied with Lister, performed the first brain surgery. Previously, surgeons had done only "skull surgery" without operating on the brain itself. Not surprisingly, the idea of brain surgery met with some reluctance. Macewen's first operation didn't take place until the patient had died. He had diagnosed a cerebral abscess and proposed removing it surgically, but the family wouldn't allow him to operate. He was given permission after the patient died and found the abscess he had expected and removed it. In 1879, Macewen had his chance on a live patient and evacuated a subdural hematoma (a collection of blood on the surface of the brain). He later published a book documenting a decade of surgery in which he claimed to have cured 63 of 74 patients with intracranial infections.

IQ Points
Toward the end of the century Americans were also getting into the brain. In 1888, William Keene Jr. became the first American to remove a benign brain tumor.

Restricted Area

In 1885, Paul Ehrlich (1854–1915) discovered that when a blue dye was injected into the bloodstream, the tissues of every part of the body except the brain and spinal cord turned blue. Ehrlich did not realize the full significance of his discovery, but later scientists would establish that a mechanism existed that prevented material from the blood from entering the brain. Ehrlich's finding was the first suggestion of the existence of what is now called the blood-brain barrier.

The idea of a person with the ability to see through buildings (Superman's x-ray vision) came from one of the last (and very definitely not least) discoveries of the nineteenth century, the x-ray. In 1895, German physicist William Roentgen was performing a series of experiments when he put his hand in front of a beam of rays coming from a cathode tube and could see the bones. What was happening was that electromagnetic waves produced dark images when reflected against objects with little density whereas parts of the body, such as bones, which are radio dense, were highlighted in white. Imagine Roentgen's excitement at seeing something no one else had ever seen or knew could be seen in this way. Because he didn't know how to explain it, he called the beams "x-rays."

Roentgen received the first Nobel Prize for Physics in 1901. He was so shy and reluctant to talk about his discovery that he snuck out of Sweden to avoid giving the Nobel lecture. Contrary to what you might expect, Roentgen didn't make a fortune with his discovery and died in poverty.

Like many other discoveries, the x-ray was not developed to understand the brain or treat it, but it has proven to be an invaluable tool from almost the day it was discovered to the present in the diagnosis of head injuries involving skull fractures. It does not help, however, in examining the brain itself because the brain's density is not great enough to show up on an x-ray. The x-ray also is the forerunner to the more advanced scanning techniques used today.

The Century's Crazy End

We've shot through about three centuries of medical advancements in a handful of pages. As the nineteenth century came to an end, the study of the physical attributes of the brain was increasingly complemented by the study of the related mental conditions.

In 1885, French neuropsychiatrist Gilles de la Tourette (1857–1904) described several movement disorders and ultimately became the namesake for the disorder that is characterized by tics and often inappropriate speech, in particular profanity. Ironically, Tourette was murdered by one of his patients who claimed he'd hypnotized her against her will and caused her to go insane.

The study of physical and mental illnesses came together in 1887 when Russian psychiatrist Sergei Korsakoff (1853–1900) described the symptoms characteristic of alcoholism. That was also a landmark year in the history of medicine in the United States because that was the year the National Institutes of Health was established.

> **IQ Points**
>
> In 1898, the Bayer Drug Company marketed heroin as a non-addicting cough medicine. A year later, the company found a much more successful product when it introduced aspirin to the public.

About the same time, Sir Francis Galton (1822–1911) was introducing the idea that intelligence is an inherited trait. He also introduced the first scientific tests to try to measure intelligence. The tests included ways to evaluate breathing capacity, the ability to detect differences in colors, smells, and weights, as well as the highest and lowest sounds a person could hear. Galton worked out of a lab in the South Kensington Museum in London, where, for a small fee, visitors could have themselves tested. Though many of the traits he tested were later shown to have little or nothing to do with intelligence, the statistical techniques he developed and the notion that intelligence could be measured were important advances that remain valid today.

Galton also went beyond describing and measuring inherited traits and suggested that it was possible and desirable to improve inherited characteristics. For example, he wanted to check the birthrate of the "unfit" and promote the reproduction of people who had higher intelligence. He called the study of how to improve the human race, "eugenics." The idea that some people were "fit" to live and others "unfit" has very serious ethical implications, and has been misused over the years by people like the Nazis who believed in eugenics to create a "master race." On the other hand, current genetic research is aimed at what some people see as the related goal of manipulating genes to create children with more positive traits (e.g., intelligence, height, strength) and fewer negative ones (e.g., susceptibility to disease).

In 1899, German psychiatrist Emil Kraepelin (1856–1926) distinguished between the mental disorders now known as schizophrenia and manic-depressive disorder. He also was the first to develop a classification of mental illnesses. From this point on, the study of mental disorders would for the first time in history receive almost as much attention as physical illnesses.

The Least You Need to Know

◆ The philosopher Descartes introduced the question of the relationship between mind and body. He believed they were separate.

◆ The invention of the microscope allowed scientists to see for the first time many of the elemental building blocks of the human body that are critical to understanding the brain.

◆ Phrenology and mesmerism were two pseudoscientific attempts to explain behavior. The former was proven to be specious, but the latter gave rise to the legitimate use of hypnotism to treat ailments.

◆ Neurons in the brain transmit information to the motor nerves, and information from sensory nerves is conveyed to the brain. The neurons are separated by gaps that electrical impulses must cross.

◆ The abilities to speak and understand language are traced to particular areas of the brain.

Chapter 4

"Modern Thinking"

In This Chapter

- ◆ Psychoanalyze this
- ◆ Get ink blotto
- ◆ Meet brain surgeons
- ◆ Scope and scan
- ◆ Lop and shock
- ◆ Relive the decade of the brain

A host of limitations impeded medical research prior to the twentieth century. Religious doctrine and politics often stifled investigation and created serious disincentives to theorists who challenged the views of the theologians. An incomplete understanding of the causes of infection made it difficult to operate without killing the patients and therefore slowed progress in surgical techniques. Legal and moral restraints on the dissection of humans hampered the study of anatomy. The undeveloped state of medical technology limited scientists' knowledge of the form and function of the brain. In the course of the next 100 years, however, these obstacles were almost entirely removed, which fostered an exponential advancement in our understanding of the brain and the treatment of illnesses affecting it.

Sex, Dreams, and Psychoanalysis

The twentieth century began with a work of monumental significance to the scientific community, *The Interpretation of Dreams* by Sigmund Freud (1856–1939), which was published in 1900. Freud began his medical career as a neurologist and became a specialist in nervous disorders. He gradually became convinced that many of his patients did not have physical illnesses but were suffering from mental conflicts that they were not even aware of.

His early work focused on identifying traumatic experiences that he believed were the cause of his patient's problems. He used hypnosis (remember our old friend Anton Mesmer from Chapter 3) to induce his patients to recall and often act out these experiences, which had the effect of helping them get over the events that had troubled them. Freud called the mental process by which these traumatic experiences were buried in the unconscious repression and described repression as one of the defense mechanisms developed by the mind to keep unpleasant thoughts beyond a person's awareness. The method used by Freud, and the school of thought he founded, is called psychoanalysis.

Freud tried to get to the unconscious and find out what was there to show how it affected behavior. He believed that dreams were a window to the unconscious mind and that they could be analyzed to determine what drives people. Dream analysis led him to the view that adult behavior is motivated, in part, by the *Oedipus* and *Electra complexes* which are the physical and emotional attachments children have for the parent of the opposite sex and the hostility felt toward the parent of the same sex.

Words of Wisdom

Although the **Oedipus complex,** which relates to boys' unconscious attraction to their mothers and hostility toward their fathers, gets most of the ink, Freud also believed that girls had similar sexual attractions toward their fathers. Freud argued that girls are envious of their father's penis and want to possess it so strongly that they dream of bearing his children, something he called the **Electra complex**. (Electra was the daughter of Agamemnon and Clytemnestra who wanted her brother to avenge their father's death by killing their mother.) This "penis envy" leads to resentment toward the mother, who the girls believe caused their castration. Not too many people buy these ideas anymore.

The sex drive plays a big part in Freud's theories. He believed, for example, that we all go through three stages—oral, anal, and phallic—in which we associate pleasure with the stimulation of certain parts of our bodies. What happens during those stages, Freud argued, could influence adult behavior.

Freud's Stages of Psychological Development

Stage	Description
Oral	In the first year of life, the mouth is the center of pleasure. If an infant receives enough mouth stimulation, it will move on to the next stage.
Anal	From ages one to three children begin to learn about their bodily functions. People who are very rigid are sometimes referred to as "anal-retentive," which comes from Freud's idea that people can get stuck in this stage where the focus is on self-control and personal hygiene.
Phallic	From ages three to six children begin to focus on their own sexual organs. This stage is also when the Oedipus and Electra complexes kick in.

One other important element of Freud's thinking is the idea that a battle is constantly going on inside our minds between the part that seeks immediate gratification—what he called the id—and our conscience, which attempts to tame these urges (the superego). The ego makes the final decision by balancing what the superego thinks is right with what the id desires.

There you have it, Freud's theory in a nutshell (excuse the pun). His ideas have become less popular over the years as new psychological theories of behavior have emerged. Nevertheless, Woody Allen and many other people still find psychoanalysis an aid to understanding and coping with their lives.

Mind Matters

In the latter half of the nineteenth century, scientists became more rigorous in their approach to investigating the brain. They began to conduct more carefully controlled experiments and documented their findings with greater attention to detail. By the early twentieth century, the scientific method gained acceptance as the predominant approach for the study of brain and behavior and the influence of folk-wisdom, speculation, and religion declines.

Testing Intelligence

In Chapter 3, we noted that Francis Galton began to develop measurements of intelligence. Galton's focus was on physical traits and responses to stimuli that were relatively easy to document. In 1904 French psychologist Alfred Binet (1857–1911) was appointed to a commission that wanted to create a system for identifying retarded children who will need special education programs. Binet disagreed with Galton's view that intelligence was related to the ability to distinguish sensory information and maintained that more complex processes such as memory, imagination, and comprehension were better indications of mental ability.

Binet and a physician named Theodore Simon (1873–1961) designed a test that could specifically distinguish between children who learned at a normal rate and those who were slow. The tests were cleverly designed so that they did not include school-related skills to insure the evaluation would not be tainted by the amount or type of schooling the child had. The scale they designed was based on a comparison between the "mental age" as scored on the test and the child's chronological age. If the mental age was lower than the chronological age by two or more years, Binet said it indicated the child was mentally "retarded." At the same time, Binet also warned against the misuse of the tests.

Defining a Disease

Alois Alzheimer (1864–1915) was a professor of psychology in Germany who worked with neurologist Franz Nissl (1860–1919). Together they published a monumental six-volume study on the cerebral cortex. At a meeting in 1907, Alzheimer presented the case of a 51-year-old woman whose symptoms included depression, hallucinations, and dementia. After she died, he found that the number of cells in her cerebral cortex was abnormally low.

Alzheimer described the condition in this way:

> The Disease begins insidiously with mild weakness, headaches, dizziness, and sleep-lessness. Later, severe irritability and loss of memory develop. Patients complain bitterly of their symptoms … Increasing loss of memory and progressive clouding of mind appear later, with sudden mood changes … the Disease leads to stupor and childlike behavior.

Famed German psychiatrist Emil Kraeplin (who was discussed in Chapter 3) proposed naming the condition after Alzheimer.

Tracking Chemical Messages

About the time Golgi and Cajal were receiving the Nobel Prize for their study of the nervous system, a group of other researchers, including Joseph Erlanger (1874–1965) and Herbert Gasser (1888–1963) who later shared the Nobel Prize in medicine, discovered that the electrical pulses within neurons caused chemicals to be released, the function of which was to send a message to other neurons using the connections between them. After a neuron "fired" its message, scientists found it took only one-thousandth of a second for it to recharge.

In 1906, Charles Scott Sherrington (1852–1932) published "The Integrative Action of the Nervous System," which described the synapse and the motor cortex. He also proposed the "million-fold democracy" theory, which suggested that the brain encodes an idea, movement, or sensation through the "voting" of many neurons.

Realizing a Nobel Dream

On Easter Sunday in 1921, Otto Loewi (1873–1961) awoke in the middle of the night with the idea for an experiment. He made a note for himself and went back to sleep. The next morning, he couldn't read what he had written. That evening, he woke up at the same time he had the night before and remembered what he had written. This time he went immediately to his lab and conducted an experiment, which ultimately proved that one of the ways nerve impulses are transmitted is chemically.

Loewi won the Nobel Prize in 1936. Two years later, the Nazis forced him to trade his share of the prize money for his life. Penniless, Loewi left his native Austria and ultimately made his way to the United States.

Testing, One, Two, Three

While working in a psychiatric hospital with adolescents, Hermann Rorschach (1884–1922) noticed that certain children gave characteristically different answers to a popular game known as blotto. From that observation he developed a series of 10 inkblots to show patients and ask what they saw in them. He believed these pictures could elicit subconscious thoughts that would provide clues to a patient's personality. In 1921, Rorschach's inkblots were published in a monograph, *Psychodiagnostik*.

IQ Points
Unless you've taken the Rorschach test, chances are you've never seen the real Rorschach inkblots. They are kept secret so that patients' reactions to the inkblots will be spontaneous. Patients also are not supposed to know how the test is administered or interpreted to make it more difficult to manipulate the test. There are no fixed correct answers, but the test would be useless if there were not certain expected responses, with those that are unusual suggesting there may be a problem. In the case of some of the inkblots, a specific answer is interpreted in a particular way. For example, whether you see a male or female in one of the inkblots is supposed to determine sexual preference. If you wanted the psychologist to think you were a heterosexual, it would help to know the answer that yields that conclusion.

Rorschach died tragically of complications from appendicitis at the age of 37, not long after introducing his test. Without his guidance, different psychologists developed their own ways to administer, score, and interpret the test. Throughout the 1940s and 1950s, the Rorschach was the test of choice in clinical psychology. It could be used, for example, to explore the fantasy life of patients without asking them about it directly, and by repeating the test, the analyst could monitor the patient's progress.

The Rorschach came under fire from critics who questioned the test's reliability and validity. Still, the test has been perhaps the most frequently written about subject in the field of psychology and has probably been more widely administered than any other psychological exam.

Creating Head Gear

In 1908, Victor Horsley (1857–1916) and Robert Henry Clarke designed the first stereotaxic instrument. This steel contraption, which looks like a scaffold for the head, may seem like something the Inquisitors would have used to torture people, but it is actually extremely useful.

> **Gray Matter**
>
> Clarke suggested to Horsley that the stereotactic method could be useful in human neurosurgery, but Horsley apparently disagreed, and the friction over the idea reportedly ended their professional relationship. Clarke subsequently submitted a patent application for a human stereotactic instrument.

Initially, scientists used the instrument to study the brain structures of cats and monkeys. It was not used for human surgery until 1947. Now surgeons use the device to stabilize a patient's head during surgery; it also has needle holders that allow doctors to precisely insert needles or electrodes in the brain. This instrument made it possible for Harvey Cushing, the father of American neurosurgery, to conduct the first experiments involving the electrical stimulation of the human sensory cortex.

I'm a Surgeon, Neurosurgeon

The work of two men, Harvey Cushing and Walter Dandy, was largely responsible for establishing the field of neurosurgery. During a time when few people owned cars and air travel was unknown, Harvey Cushing (1869–1939) succeeded in revolutionizing the treatment of brain injuries. When he started, neurosurgery was usually an act of last resort and almost always fatal, but Cushing helped to make the practice a common and often successful method of treatment. He was perhaps the first person to be referred to as a brain surgeon.

During his lifetime, Walter Dandy (1886–1946) was one of the most renowned neurosurgeons in the world. Dandy did more neurosurgical operations, wrote more papers and books, and contributed more knowledge and diagnostic tests and new ideas to neurosurgery than perhaps anyone other than his teacher, Harvey Cushing. He also set out to perform operations that Cushing said couldn't be done, seeking new approaches to minimize the very high risks then associated with the still new practice of neurosurgery. The work of Cushing and Dandy did much to advance the treatment of brain injury and disease.

The Father of Neurosurgery

Harvey Cushing performed some of the earliest operations on brain tumors in the United States. By his own admission, most of his early efforts were unsuccessful, with as many as

90 percent of his patients dying. Still, he introduced a number of important innovations, such as the use of tourniquets to stop the scalp from bleeding and his invention and production of clips to tie off arteries.

In 1910, Cushing successfully removed a tumor from General Leonard Wood, an influential person of the time, which helped establish the surgeon's reputation. Cushing's status continued to grow, especially after Harvard University asked him to help in the planning and design of a new hospital. Cushing served as the chief surgeon in the new Peter Bent Brigham Hospital and produced an astounding amount of research and clinical studies over two decades before retiring in 1932.

In 1925, Cushing published a two-volume biography of his mentor, Sir William Osler, another of the giants in the history of medicine. *The Life of William Osler* was awarded the 1926 Pulitzer Prize for biography.

One of Cushing's most important contributions to the study of the brain was the creation of a Brain Tumor Registry. This astounding collection contained more than 2,000 case studies, including brain specimens, 15,000 photographic negatives, microscopic slides, thousands of pages of hospital records, and notes from a career that spanned from the 1880s to 1936.

> **IQ Points**
>
> Cushing had such a high opinion of himself that he left money in his will for the writing of his biography.

Cushing came out of retirement to take a position at Yale where he was a professor of neurology and also director of studies in the history of medicine. Cushing donated his work to the school and persuaded several of his friends to do the same, which formed the basis for what came to be one of the world's largest collections of material on medical history.

Though Cushing gets most of the credit, his tumor registry would not have been created if not for the brilliance and dedication of Dr. Louise Eisenhardt. She began as his editorial assistant and later went to medical school herself where she graduated with the highest grades ever awarded at Tufts. She worked for Cushing during school and continued to do so after graduating as the first female neuropathologist in 1925. Eisenhardt kept records of all Cushing's procedures and didn't let him see it for fear he might be tempted to try to make his statistics look better. He finally got to see her journal on his seventieth birthday. Eisenhardt also made her own mark as the first editor of the *Journal of Neurosurgery*, a post she held for more than 20 years.

A Dandy of a Surgeon

Walter Dandy served as a resident under Harvey Cushing at Johns Hopkins. When Harvard offered Cushing a job to be the chief surgeon at a new hospital, Dandy expected to follow his mentor. However, Cushing didn't take Dandy with him, and the young physician suddenly found himself without a job. Dandy ended up working in a laboratory, which was fortuitous because the job enabled him to make a number of important discoveries. Meanwhile, he and Cushing became lifelong antagonists.

One of Dandy's innovations was a "bloodless" surgical approach to the base of the brain. Dandy also developed a new method of removing brain tumors. He had found that most tumors couldn't be removed because they could not be distinguished from the surrounding brain tissue. Dandy adopted the radical approach of removing the lobe containing the tissue. This surgery was done only when the patient would have died without it.

Dandy made three significant contributions to medical history. In 1918, he introduced ventriculography, a test during which air is pumped directly into the ventricles of the brain. This procedure is followed by an x-ray examination of the skull. The technique made it possible for the first time to accurately diagnose the location of intracranial brain tumors. A year later, Dandy invented another test, pneumoencephalography, which involves injecting air into the ventricles of the brain through a spinal puncture. These tests formed the basis of neurological imaging for the next 50 years. Finally, he demonstrated the circulation of cerebrospinal fluid.

> **CAUTION**
>
> ## Code Blue
>
> Given the change in personality caused by the damage to the brain of Phineas Gage (the man who had a metal rod rammed through his head and survived, as discussed in Chapter 3), it was feared that radical surgery such as that performed by Dandy would also create severe behavioral changes. In fact, the impact was found to be minimal.

Controversial Cures

Today, when we think about the treatment of mental illness, the first things that come to mind are probably therapy and medication. If we asked you to describe how you thought the mentally ill were treated in the old days, we suspect you'd think of harsh methods that were somehow less humane. Though you may not know anyone who has undergone such treatment, popular media has introduced us all to the once common practices of lobotomy and shock therapy.

Off with Their Lobes

In 1927, Portuguese neurologist Egas Moniz was the first to use cerebral angiography, a method he developed for studying arteries in the brain. He is better known, however, for his discovery in 1936 that he could reduce the anxiety of monkeys by cutting the nerve fibers in the frontal lobes of their brains. The operation, which became known as a lobotomy, had a similar impact on humans. The procedure altered behavior, usually reducing the intensity of the emotion in disorders such as anxiety, neurosis, severe depression, and manic-depressive psychosis. Anxieties, thoughts, or delusions that distressed and incapacitated the patient sometimes persisted, but they became more remote and of little or no concern. Unfortunately, other matters such as household duties, sexual propriety, or regard for the feelings of others sometimes also became of no interest to the patient.

In medicine, the cure is sometimes worse than the disease. This was the case for a form of lobotomy once used to treat epilepsy. In 1953, a patient who suffered severe epileptic seizures was cured by the removal of the middle of the temporal lobe of his brain. Unfortunately, the operation permanently damaged his memory. He could recall events for a short period before the surgery, but he could not remember anything afterward. Those of us with bad memories may joke that we can't remember what we had for breakfast, but this poor fellow really couldn't. "There are no yesterdays" is the way he described his condition.

The only positive result of this operation was that its negative effects were publicized enough to ensure that it was not repeated. Currently, partial lobotomy and even hemi-spherectomy (the removal of half the cerebrum) is routinely performed in order to treat epilepsy.

> **IQ Points**
>
> In the United States alone, roughly 35,000 people had lobotomies in the 40 years following Moniz's discovery as it became a popular treatment for severe mental illnesses. When the severity of the side effects became clear, however, it fell out of favor, and now the operation is rarely performed.

A Shocking Treatment

The term "shock treatments" usually conjures up the image of Jack Nicholson's character in *One Flew Over the Cuckoo's Nest* being strapped down with electrodes on his head and then jolted with enough electricity to turn him into a vegetable. Yet the first shock treatments did not involve electricity at all. In 1917, Julius von Wagner-Jauregg (1857–1940), a psychiatrist with a particular interest in the legal rights and the treatment of the insane, began experimenting with inoculations of tuberculin to induce fever. He hoped the "shock" of the fever would cure mental illness. It did not.

Though Wagner-Jauregg's treatment of the insane failed, his approach did yield other positive results. He cured people suffering with general paresis (a disabling, common, and previously incurable manifestation of tertiary syphilis) by inoculating them with malarial parasites. The advent of penicillin made this therapy obsolete. Nevertheless, Wagner-Jauregg was awarded the Nobel Prize for his work, which held out the hope of finding other physical treatments for mental illness. (On the other side of the ledger, Wagner-Jauregg was an anti-Semite who joined the Nazi Party and supported the crusade to prevent criminals and the mentally ill from procreating.)

> **Gray Matter**
>
> In 1946 President Truman signed the Mental Health Act, which awarded grants to establish mental health clinics and treatment centers. The National Institutes for Mental Health (NIMH) were also established at this time.

In 1938, Ugo Cerletti and Lucino Bini used the "real" electroshock on the first human patient. This treatment involves the use of an electric current that is generated through two electrodes placed over the temples. The shock causes unconsciousness and a convulsive seizure. After treatment patients experience some memory loss, but it usually is not permanent.

The development of powerful tranquilizing drugs combined with the negative image of the treatment led to a decline in its use. Nevertheless, it remains among the more effective techniques used in severe cases of various mental disorders, including depression and schizophrenia.

New Tools for the Trade

Considering that we are dealing with such a sophisticated and delicate organ, it may be somewhat surprising that research and surgery to this point have been so low-tech. Think about it, most of what has been involved with brain surgery was a device to make a hole in the skull, something we saw that people did thousands of years ago, a knife, and a few other relatively primitive implements. Sure there were some useful innovations, such as the stereotaxic instruments we described earlier in this chapter, but most of the advances have been more in terms of the approaches to studying the brain. During the course of the twentieth century, particularly the latter part, new technologies began to revolutionize the study of the brain as well as the diagnosis and treatment of physical problems. Doctors still continued to do their share of poking and prodding, but more and more information could be gleaned from machines capable of taking different types of pictures of our insides and computers that could interpret those images.

Current Events

In 1929 Hans Berger (1873–1941) placed electrodes on the scalp and recorded the electrical currents in the brain. The recording device he used could detect brain waves even when the patient was conscious. This kind of recording later became known as the electroencephalogram. Berger, who was a psychiatrist, believed the currents he recorded were a result of psychological energy.

A Closer Look

The physics of light limited light microscopes to 500x or 1,000x magnification. Although this magnification was sufficient to see the detail of tiny elements of the body, scientists knew that there was even more to see. In 1931 Ernst Ruska (1906–1988) developed the electron microscope in Germany. This microscope used beams of electrons instead of light to show the specimen at magnifications exceeding 50,000x.

Newer types of electron microscopes are even more powerful and are capable of magnifications of 300,000x or more. In 1981, Gerd Karl Binnig (born 1947) and Heinrich Rohrer (born 1933) invented the scanning tunneling microscope (STM), which uses an entirely different method to magnify the specimen and provided the first images of individual atoms on the surfaces of materials. For this invention, Binning and Rohrer shared the 1986 Nobel Prize in physics with Ernst Ruska.

A Homunculus Brain

Over the course of about two decades starting in the 1930s, Canadian neurosurgeon Wilder Penfield (1891–1976) made a major breakthrough in the understanding of the differences in function of different parts of the brain when he obtained the consent of patients to perform operations while they were conscious. Penfield used electric probes to stimulate the cortex of the brain and then asked the patients what they experienced.

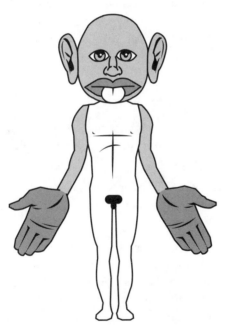

The homunculus shows the size of the parts of the body based on the amount of cortex devoted to their control.

Simple Representation of the Sensory Cortex

By documenting his patients' responses, Penfield created maps of the brain. They were not ordinary maps, however, but rather were pictures of "little men" (homunculi) that depicted the size of body parts according to the amount of the cortex devoted to controlling them. Thus, a homunculus would have large feet, very large hands, and big lips because relatively greater amounts of the cortex are devoted to controlling those parts of the body than others.

The Scan Man Can

Another diagnostic milestone occurred in 1972 when Godfrey Hounsfield (born in 1919) developed computerized axial tomography, better known as the CAT scan (aren't you thankful for acronyms?). In this somewhat unpleasant, but painless, procedure, the patient puts his or her head in a cylinder, and x-rays are taken of the brain from different angles. Instead of being recorded on a photographic plate, as in a conventional x-ray, the images are fed into a computer, which puts the information from all the x-rays together to create a cross-sectional image that has 100 times greater detail than an ordinary x-ray. These pictures make it possible to identify physical problems such as tumors and lesions. They do not provide any information, however, about the operation of the brain.

Two years after the invention of the CAT scan, a new technology was developed. The technical name for it is positron emission tomography, but most people just call it the PET. This procedure is slightly more involved than the CAT scan and requires the injection of glucose or water with a radioactive component that accumulates in parts of the brain depending on how hard they are working. The radioactive elements show up in the PET images and allow doctors and researchers to see which parts of the brain are active during particular tasks.

Not for the Claustrophobic

The third major development in scanning technology was magnetic resonance imaging (MRI), which was used on a human for the first time in 1977. Raymond Damadian, Larry Minkoff, and Michael Goldsmith invented the MRI.

The first MRI machine was named "Indomitable" because it was the culmination of their seven-year effort to accomplish what many said was impossible. That original machine is now in the Smithsonian Institution in Washington.

The MRI is a powerful diagnostic tool; it offers even greater detail than other imaging techniques and enables the physician to tailor the scan to a particular problem. The MRI uses a giant magnet to line up the protons in the body and align them with the north-south polarity of the magnet. (The magnetic field created in the process is 30,000 times stronger than the Earth's magnetic field.) As with the CAT scan, the resulting information is fed into a computer, which can quickly translate it into images from a variety of angles and directions.

The MRI exam is painless, but it can be unpleasant and even provoke anxiety in the claustrophobic. The MRI patient lies down on a table that is slid into a coffin-like cylinder The procedure typically takes 20 to 30 minutes, and the banging noise of the machine is so loud that the patient is given earplugs, which dampens but does not eliminate the pounding noise. The good news is that doctors usually will prescribe a mild relaxant to help calm the patient's nerves. Another drawback to the MRI is that it is very expensive and, therefore, not available in many small hospitals.

Yet another new technology is magnetoence-phalography (MEG for short), which measures the magnetic field generated by the differential electrical activity of the brain. The information provided by MEG differs from both CAT and MRI scans. The latter two provide structural and anatomical information while MEG can provide images of neurological functions to measure the electrical activity of the brain in real time.

> **IQ Points**
>
> The MRI was originally known as the NMR for nuclear magnetic resonance. The name was changed because of the concern that patients would be afraid of getting into something nuclear.

Your Own Painkiller

In the early 1970s, researchers discovered a previously unknown substance in the brain called enkephalin. They found that this chemical worked much like morphine to ease and sometimes eliminate pain. It does so by interfering with the transmission of the pain impulse from the nerve cell.

In stressful situations, the brain produces more enkephalin, which can sometimes mask an injury that occurs when someone is anxious. Thus, for example, athletes may not feel the pain of an injury during a game, but will suddenly realize they've been hurt after the game is over and their stress levels have diminished. Scientists also found that certain drugs block the release of enkephalin, and this block can cause people to suffer greater pain.

Another important chemical discovery was the discovery of the existence of hormone-like chemicals called *prostaglandins*. These chemicals help your body transmit pain signals, and thus help you feel pain. The British pharmacologist John Robert Vane (born in 1927) showed in 1971 that aspirin's effectiveness is related to its ability to block the production of certain prostaglandins.

> **Words of Wisdom**
>
> The name **prostaglandin** comes from the fact that the first prostaglandin was found in semen and was thought to originate from the prostate gland.

Decade of the Brain

President George Bush declared the decade starting in 1990 the "Decade of the Brain." This declaration helped generate a burst of research around the world that led to numerous exciting discoveries and advances.

Thanks to the improvement in microscopes, many of the greatest advances were made at the molecular level. The breakthroughs in genetics allowed researchers to identify the genes that control specific functions in the brain and many of the mutations that cause diseases such as Parkinson's, Huntington's, and Alzheimer's.

Technology is a key to many of the recent medical advances. MRI, PET, CAT, and other imaging methods have aided the study of both healthy and sick brains. Some long-held beliefs have been shattered as a result of this study, such as the notion that specific areas of the brain control particular functions. We now know that a single function may involve a number of different parts of the brain. For example, the processing of visual information was once believed to be largely controlled by one area of the brain, but it is now clear that dozens of areas are involved.

Scientists also gained a better understanding of the genetic basis for the development of the brain's neural circuitry. Other research demonstrated how behavior can alter the brain's circuits over time. For example, in the case of someone who is born deaf and uses sign language rather than spoken language, the visual center of the brain takes over many of the functions ordinarily performed by the auditory system.

The idea of transplanting a brain remains largely restricted to science fiction; nevertheless, the 1990s did witness advances toward the ability to regenerate or transplant nerves. This research holds great promise for the treatment of a variety of diseases.

In medicine, as in most scientific fields, the advancements in the twentieth century were exponential and revolutionary. In the specific case of brain research, the amount of knowledge of the anatomy and physiology of the brain during those 100 years was greater than all of the accumulated knowledge of the previous 6,000. It is only in the last 50 years or so that we have begun to learn what the different parts of the brain do. Still, even after the "Decade of the Brain," we have entered a new millennium with many old questions still unanswered and perhaps even more new ones.

The Least You Need to Know

- Sigmund Freud changed the way we think about human behavior by attributing our actions to conflicts within our subconscious.
- By the 1930s, researchers had discovered that electrical nerve impulses are transmitted chemically.
- Harvey Cushing and Walter Dandy were the preeminent brain surgeons of their day, performing thousands of procedures and introducing important innovations that made surgery a more effective means of treating brain diseases.
- Technological advances improving the magnification of microscopes and the ability to generate computer-generated images of the body revolutionized the study and treatment of the brain.
- President Bush declared the 1990s to be the "Decade of the Brain," which stimulated a period of intense research and a host of advances in the understanding of the brain, but left many questions unanswered and introduced new mysteries.

Part 2

Anatomy 101

This part of the book provides a very quick overview of the anatomy of the brain to help you understand the different parts and what they do. There won't be a test, but learning these major anatomical features of the brain will be useful because we will be referring to them throughout the remainder of the book.

Some of this information is a bit technical (even med students find it tough), but we're confident you can handle it. If some of the words are unfamiliar or hard to pronounce, take comfort in knowing that at least one of the authors felt the same way you do. If nothing else, you'll beef up your vocabulary a bit and impress your friends with nifty tricks such as naming all the cranial nerves (or at least reciting the mnemonic that helps med students remember them).

Chapter 5

The Commander-in-Chief

In This Chapter

- Brain protection
- Two halves make a brain
- Those laboring lobes
- The body's thermostat

The cerebrum is the largest piece of the brain, comprising 85 percent of its weight. When you speak, listen to music, watch television, feel heat or cold, jog around the block, write a poem, recall a phone number, or read these words, some part of your cerebrum is involved. The cerebrum even controls actions that you don't think about, such as shivering, breathing, and digesting food. In short, the cerebrum is in charge of you, and its highly developed nature is what separates us from other animals.

To gain an appreciation for this all-important piece of anatomy, you'll have to put it to use by reading the rest of this chapter.

Your Hard Hat and Other Protective Devices

Before getting into the nitty-gritty description of your brain, we should at least pay a paragraph or two of tribute to the part of the body that protects this vital piece of equipment: your skull. If you didn't have a bony shield surrounding the soft tissue of the brain, it would be easily damaged. (After all, your brain is less than a half inch from the hair on your head.)

Quick, how many bones are in the skull?

If you answered 22, go to the head of the class (excuse the pun). You get bonus points if you knew that each ear contains an additional three bones, which don't count as part of the skull.

Eight bones surround your brain: one frontal bone, two parietal bones, two temporal bones, one occipital bone, one sphenoid bone, and one ethmoid bone. Together they are called the cranium. (The 14 bones in the face are also considered part of the skull.) The jagged lines where the bones of the cranium connect are called sutures.

The skull is not the brain's only protection, however.

Just inside the cranium are three protective membranes that cover the brain and act as a shock absorber. These membranes are called meninges. The inner surface of the skull is lined with a tough layer, the dura matter. A thinner layer, the arachnoid, which takes its name from its resemblance to a spiderweb, is just inside the dura matter. Finally, a more delicate inner layer, the pia matter, directly sheaths the brain. The meninges follow the contours of the brain and contain vessels that carry blood to and from the surface of the brain.

The brain's cover.

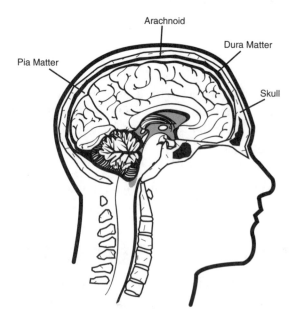

Cavities in Your Head

The brain is further protected by a clear, watery fluid called cerebrospinal fluid (CSF) that forms a thin cushioning layer between the soft tissues of the brain and the hard bones of the skull. It acts like those annoying polystyrene peanuts that are used in packaging to protect goods from being damaged in transit. Half a cupful of CSF is enough to keep the brain insulated. The fluid also gives the brain buoyancy to help reduce pressure inside the head.

The cerebrospinal fluid that flows around the outside of the brain is produced by the four holes in the brain (yes, you do have holes in your head) known as *ventricles*. The fluid in the ventricles prevents the brain from collapsing under its own weight and helps to deliver nutrients in particular hormones and expel waste products. The CSF circulates through the ventricles and around the brain and exits the fourth ventricle through an opening to the spinal cord called the *foramen magendie*. (Curiously enough two ventricles are referred to as the third and fourth ventricle, but the others are not called one and two. Instead they are referred to as the lateral ventricles.) Eventually, the CSF is reabsorbed over the surface of the brain into large veins that carry the fluid back to the heart.

The ventricles.

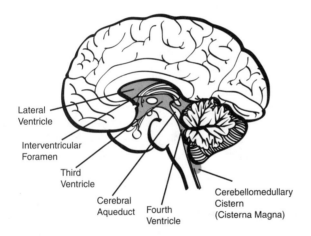

Lateral
Ventricle

Interventricular
Foramen

Third
Ventricle

Cerebral
Aqueduct   Fourth
Ventricle

Cerebellomedullary
Cistern
(Cisterna Magna)

The examination of CSF is often helpful in diagnosing diseases of the central nervous system because the CSF makeup is a reflection of what's going on in the brain. If you can measure the number of red and white blood cells, protein content, the sugar content, chemical content, and the bacteriology of the CSF, it makes it easier to diagnose diseases such as meningitis.

Under certain conditions, the ventricles may overproduce cerebrospinal fluid, the body may not absorb the CSF, or some obstruction may interfere with the ventricular system. Any of these problems can cause hydrocephalus, sometimes called "water on the brain," a potentially fatal illness characterized by the enlargement of the skull. (We'll have more to say about this ailment in Part 5.)

The Blood-Brain Barrier

The brain also has a defense system known as the blood-brain barrier. Even though the brain gives itself top priority in the blood distribution system and must receive a constant supply of blood bringing food and oxygen to keep it working properly, the flow of blood must be regulated to prevent large molecules carried in the bloodstream from damaging fragile brain tissue.

CAUTION Code Blue

Without enough blood in your brain, you become faint, and if enough time passes (only about three to five minutes), brain cells start to die. This condition can cause brain damage and ultimately death.

In other parts of the body, cells travel through spaces in the walls of the *capillaries*. In the brain, the walls of the capillaries have much tighter spaces to further restrict the type of cells that can fit through.

The blood-brain barrier is a complex system. For example, the brain does need some large molecules for nutrition that won't fit through the gaps in the capillary walls. To compensate, the body produces enzymes and performs some nifty chemical tricks that allow only the needed molecules to pass through the capillary walls.

Words of Wisdom

The word **capillary** comes from the French or Latin word for hair. Capillaries are microscopic blood vessels that form a network of tiny tubes throughout the body, connecting the smallest arteries and veins. Capillaries have very thin walls composed of a single layer of cells that distribute oxygen and nutrients from the blood into the body tissues and absorb waste and carbon dioxide.

Blood Supplies These to the Brain	Blood Removes These from the Brain
Oxygen	Carbon dioxide
Vitamins	Lactate
Hormones	Hormones
Fats	Ammonia
Amino acids	Carbohydrates

The Thinking Part

When you picture the brain, you probably see the basic cross-section that is in every anatomy book, such as portrayed in the following figure. The first thing that comes to mind (forgive the pun) when you think about the brain, however, is probably the large mass of tissue in the front—the gray matter, the cerebral *cortex*. Let's face it, the other structures may be interesting, and, as we'll see, indispensable to survival, but the meat of the brain is the cortex, where most of the functions associated with the mind—memory, creative thought, intelligence—are based.

Gray Matters

The exterior surface of the cerebrum is a folded layer of cell bodies known as the cerebral cortex, often called the gray matter because of its color. It is composed of about 50 billion nerve cells.

The cortex is about .16 of an inch thick, and its total surface area is about 16 square feet, about the size of an office desktop, but it is folded up so that it fits into your skull. The folds, or convolutions, are made up of ridge-like bulges known as *gyri*, which are separated by small grooves called *sulci* and fissures.

Nerve fibers connect the cortex with the cerebellum, brain stem, and spinal cord and hold parts of the cortex itself together.

The largest portion of the cortex is the association cortex. Every lobe of the brain has areas of association cortex that analyze, process, and store information. These association areas make possible all of our higher mental abilities, such as thinking, speaking, and remembering.

IQ Points
The word **cortex** comes from the Latin word for the "bark" of a tree. This name is appropriate because the cortex is a sheet of tissue that makes up the outer layer of the brain.

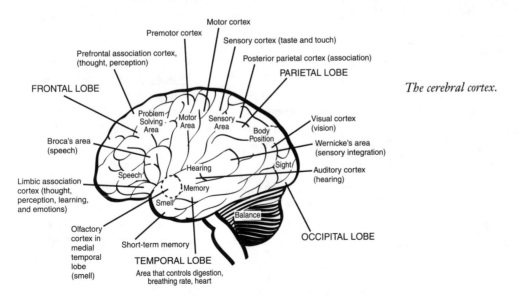

The cerebral cortex.

The Ganglia's All Here

Sit up straight!

That message should have been delivered to your brain just now. Deep within the cerebrum, large masses of nerve cells form the basal ganglia, and these structures play an important role in the control of movement and the maintenance of posture. Right now this part of your brain is involved in sending messages to the muscles in your back to pull back your shoulders and straighten your spine. Diseases that cause the loss of control of the muscles and affect coordination, such as Parkinson's disease and Huntington's chorea, affect this region of the brain, among others.

Multiple Brains

Your brain is divided into halves by a groove known as the longitudinal *fissure*. These halves are (not so) creatively named the right cerebral hemisphere and the left cerebral hemisphere and are connected by white bundles of nerve fibers. Like many of the other subparts of the brain, these bundles have a largely forgettable name, commissures. Better known is the main connection, which is called the corpus callosum. This "white matter" connects the two sides of your brain and enables the two hemispheres to share information.

It seems logical that the right hemisphere of your brain would control the right side of your body and the left hemisphere would control the left side. This idea may be logical, but it's not correct. Actually, each hemisphere controls the opposite side of the body. The nerve cells that carry messages from the brain cross over at the base of the brain.

For unknown reasons the left side of the brain is usually dominant. Because the left side controls movement on the right side of the body, most people have slightly more control and strength in their right hands and legs (which grows with use and practice). Thus, about 90 percent of the population is right-handed.

As we'll discuss in greater detail in Chapters 9 and 16, the two halves of the brain also have different responsibilities. For example, the right side is associated with creativity, recognizing objects, and solving puzzles while the left side controls speaking, writing, and logical thinking.

Words of Wisdom

A **fissure** is a slit or groove between body parts.

IQ Points

Famous artists such as painters, musicians, writers, and sculptors have been disproportionately left-handed. For example, da Vinci, Michelangelo, Picasso, Jimi Hendrix, and Mark Twain were all southpaws. Never mind all the great baseball players, from Babe Ruth to Sandy Koufax to Barry Bonds who are also left-handed. Then again, the slightly less famous authors of *Idiot's Guides* are right-handed.

The hemispheres are further divided into four regions called lobes. These lobes are separated by the central and lateral fissures and are named for the bone of the skull that covers them:

- The frontal lobe is—you guessed it—in the front of the brain.
- The parietal lobe is roughly in the middle of the brain.
- The temporal lobe is on the lower side of the brain adjacent to your ears.
- The occipital lobe is in the back of your head.

Code Blue

Surgeons sometimes treat epilepsy by removing a part of the brain. To be sure that the operation does not render the patient unable to speak, they perform the Wada Test by injecting a barbiturate into the carotid artery to knock out one part of brain. This test allows doctors to determine whether that patient's speech area is on the right or left side.

You have two of each lobe, one in each hemisphere, but the form and function of each lobe are not the same in both hemispheres. For example, the area associated with speech is found only on the left side of roughly 95 percent of right-handed individuals (and 60 to 70 percent of lefties). The following sections describe each of these lobes.

Full Frontal

The frontal lobes form the bulge in the front of the brain. Areas in the frontal lobes called the motor cortex send out nerve impulses that control the voluntary movements of all the skeletal muscles. One of the most important parts of the frontal lobes is Broca's area, which is the control center for speech. Other areas of the frontal lobes affect personality, behavior, and the senses.

Making Sense

Some areas of the cortex, called the sensory cortex, receive messages from the sense organs as well as messages of touch and temperature from throughout the body. These areas are located in the middle of the cortex, the parietal lobes. For example, the primary somato-sensory cortex is a wide strip in the parietal lobes that runs from ear to ear over the top of the brain and deals specifically with responses to the skin being touched.

You may recall the discussion in Chapter 4 of Wilder Penfield's experiments in which he stimulated the brains of patients during surgery and asked them to describe what they felt. From those descriptions Penfield developed a map of the brain and those funny drawings known as homunculi. What those little men show is that even though your arms and legs and midsection compose most of your body area, they do not require much of your cortex to control the sensations you feel there. On the other hand, the sensations in your face and hands require a significantly greater proportion of your brain. Why? Well, think about how much more sensitive your face and hands are than the rest of your body. The skin in these areas has more receptors that pick up sensations when they are touched.

Area of the Cortex	What It Does
Prefrontal cortex	Problem solving, emotion, complex thought
Primary motor cortex	Initiation of voluntary movement
Motor association cortex	Coordination of complex movement
Primary somatosensory cortex	Receives tactile information from the body
Sensory association area	Processes multisensory information
Visual association area	Complex processing of visual information
Visual cortex	Detection of simple visual stimuli
Auditory association area	Complex processing of auditory information
Auditory cortex	Detection of sound quality
Wernicke's area	Language comprehension
Broca's area	Speech production and articulation

Another subdivision of the parietal lobe is the sensory association area, which is involved in additional processing of sensory information. For example, the ability to detect the size, shape, weight, and texture of an object, even without looking at it, is made possible by this part of the brain.

Temp Work

Referred to by one writer as the hi-fi area, the temporal lobe is the center for hearing and collating audio information. When a siren blares, the sound waves enter your ear and are transmitted to the primary auditory area in the temporal lobe. Among the important areas of the temporal lobe is the one discovered by Wernicke that is involved in understanding language.

Gray Matter

The function of certain areas of the cortex is still unknown. Other brain functions, such as personality, involve the cortex, but, unlike language or vision, there is no one center that controls the behavior. The parts of the cortex that can't be associated with particular functions are called nonspecific areas.

Eyes in the Back of Your Head

Visual information transmitted from the eye is first processed in the occipital lobe. Impulses from the retina of both eyes arrive in this part of the cortex. The images are divided up so that the left half is sent to the right side of the brain and the right half of the image goes to the left. No one knows why the brain works with this type of crossover, but somehow the two halves are put together so that you see the whole image properly.

The processing of visual information not only allows you to see particular objects, but it also permits the understanding of spatial relationships and provides helpful cues for movement. For example, it helps to see where you are putting your feet when you walk or run or where an object that you are trying to grasp is.

Small but Important

That big old cortex may grab the headlines because of its size and its role in the thought process, but some of the smaller parts of the brain are what keep you alive. The thalamus and the hypothalamus connect the bulging part of the brain with the stem and handle some vital bodily functions.

The Relay Station

The egg-shaped thalamus consists of two rounded masses of gray tissue lying within the middle of the brain, under the cerebrum and between the hemispheres. The thalamus is the gateway to the cerebral cortex. All sensory information, except the sense of smell, travels through the nerves to the thalamus. After the cerebrum interprets the information, the thalamus relays the commands for movement. The thalamus also has receptors that interact with certain drugs to relieve pain.

The Regulator

The hypothalamus, a structure that weighs about two-tenths of an ounce and is about the size of two peas, is the body's thermostat. It controls most vital functions, including breathing, heartbeat, digestion, food intake, water balance, body temperature, gastric acid secretion, sleep, and glandular function. You may be thinking, "Sure, those functions are important, but they're not the most exciting aspects of human existence." Hold on, the little mass of nerves under the thalamus is also involved in emotion and sexual activity.

The hypothalamus also responds to stress. This stress can originate inside the body or outside. Examples of external stressors are the blaring of a radio, being stuck by a pin, or seeing a baseball heading for your face. Internal stress might be caused by feelings of depression or anxiety. When stress occurs, the hypothalamus sends a message to the pituitary gland, which activates the adrenal gland to secrete an increase in adrenaline, which, in turn, causes an increase in heart rate, the dilation of the pupils so you can see better, the contraction of skin blood vessels so that more blood goes to the muscles and brain, and the shutdown of digestion.

If you want to get a little more technical, the way this process works is that the corticotrophin releasing factor (CRF) from the hypothalamus stimulates the pituitary gland to secrete ACTH (adrenocorticotrophic hormone), which, in turn, stimulates the adrenal gland to secrete adrenaline. Aren't you glad we said it the other way first?

As with all the other parts of the brain, the activities of the hypothalamus take place without you being conscious of them. The hypothalamus makes you sweat when you're hot to cool the body and causes you to shiver to increase heat production when you're cold. It is the hypothalamus, not your stomach, that tells you if you are hungry or thirsty and if you've had enough to eat or drink. This little beauty also regulates your reactions to strong emotions, so when you scream, blush, or cry, it is the work of the hypothalamus.

The Least You Need to Know

- The cerebrum is the largest and most important part of the brain.
- The skull, the meninges, the cerebrospinal fluid, and the blood-brain barrier all serve to protect the brain.
- The brain is divided into the left and right hemispheres. The left hemisphere of the brain controls the right side of the body and vice versa. The dominance of the left side of the brain in most people explains why most people are right-handed.
- Each brain hemisphere is divided into lobes. The frontal lobe is the center for speech, personality, behavior, and the senses; the temporal lobe controls hearing, audio information, and understanding of language; the parietal lobe is the center for receiving sensory information; and the occipital lobe is responsible for visual information.
- The thalamus and hypothalamus are small but essential components of the brain. The former relays commands for movement and most sensory information. The latter controls most vital body functions, including breathing, heartbeat, and digestion.

Chapter 6

The Coordinator

In This Chapter

- ◆ Unfolding the cerebellum's mysteries
- ◆ Keeping your balance
- ◆ Acting without thinking
- ◆ Relaying information
- ◆ Growing from the brain stem
- ◆ Understanding glands

The cerebellum looks like two clams side by side and is about the size of a fist. Its looks can be deceiving, though; like the cerebrum, the folds on the cerebellum conceal 85 percent of its mass. The cerebellum is the second largest component of the brain, comprising a little more than 10 percent of its volume. The importance of the cerebellum is clear when you consider that more than 50 percent of the neurons in the entire central nervous system are found within it.

Lobes: They're Not Just for the Ears

The *cerebellum* is in the back of the brain, beneath the occipital lobes. Like the cerebrum, the cerebellum also has gray matter on the outside and white matter on the inside. It also has its own cortex, called the cerebellar cortex. The outer

IQ Points

One of the deep nuclei in the cerebellum is the dentate nucleus. This nucleus is the major relay station for cerebellar action on the regions of the cerebral cortex involved in motor control.

part of the cortex is the molecular layer, which has few cells. Below that layer is a layer comprised of large cells known as Purkinje cells. A third dense layer of tiny neurons called granule cells lies below that.

Purkinje cells are the most complex nerve cells and interact with perhaps as many as 100,000 other nerve fibers, making more connections than any other cells in the brain. These cells send their axons to the deep nuclei of the cerebellum to eventually reach the thalamus, midbrain, and medulla. Sensory pathways bringing information from the body to the cerebellum generally end in the Purkinje cells. Deep nuclei are way stations for specific nerve impulses.

Like the cerebrum, the cerebellum is divided into two hemispheres. Each cerebellar hemisphere consists of three lobes. This fact won't be on the test, but you can amaze and astound your friends by telling them about the flocculonodular, anterior, and posterior lobes of their cerebellum. The flocculonodular lobe is the area that receives sensory input from the ears, which is used to maintain balance. The anterior lobe gets messages from the spinal cord that let the brain know what some of the other moving parts of the body are up to. Finally, the posterior lobe communicates with the cerebrum.

Words of Wisdom

Cerebellum comes from the Latin word meaning "little brain."

Ah, the **vermis**. It's one of those romantic sounding words, isn't it? This attractive name comes from the Latin word for worm because that is what the vermis looks like. In fact, many anatomical structures are named based on their shapes and appearance because the first people to see them usually had no idea of their function.

The two hemispheres of the cerebellum are connected by a finger-like bundle of white fibers called the *vermis*. The vermis was considered quite important in early medical history because the ancient Greek physician Galen (who was discussed in Chapter 2) believed it controlled the flow of pneuma. Like many of Galen's ideas, this idea was wrong, but it was accepted for several centuries. Scientists now know that the vermis acts as a relay station between the cerebellar cortex and the spinal cord.

Words of Wisdom

The **foramen magnum** is the opening in the base of the skull through which the spinal cord exits the skull and goes into the neck.

Another portion of the cerebellum is known as the tonsils. (These are not the tonsils that become inflamed and are removed from your throat when you're a kid.) There are two cerebellar tonsils, one associated with each hemisphere. They are of no great functional significance, but they are extremely important in a number of disease states of the cerebellum because as the disease causes swelling in the cerebellar hemispheres, the tonsils are forced into the *foramen magnum*. This situation gives rise to a number of extremely important symptoms that are described in Chapters 17 and 18.

Walking a Straight Line

The cerebellum has three principle functions: the maintenance of equilibrium, the regulation of muscle tone, and the coordination of muscle movement. The cerebellum helps to maintain equilibrium through its relations with the vestibular system. The vestibular system has to do with coordination and balance. It begins with input from the inner ear that travels through the vestibular nerve into the brain. (The vestibular nerve is also called cranial nerve VIII; all 12 cranial nerves will be discussed in Chapter 8.) While preserving the normal position of the body at rest and in motion, the cerebellum also helps to produce the muscle tone that is necessary to maintain this normal position.

Even those of us who tend to be klutzes do not usually have trouble standing up straight and walking without falling over. We don't give these actions a moment's thought, but our cerebellums are working at warp speed to make sure we don't lose our balance. No conscious activity takes place in the cerebellum; it works automatically to coordinate one or more muscles to produce movement. Often compared to a computer, the cerebellum processes information that comes in from the ears, eyes, muscles, and tendons and, without storing any of the information, immediately sends messages back out to modify the position of the limbs.

How does this process work? The body has a feedback loop between the muscular system and the nervous system. You'll learn more about the role of nerves in this loop in a later chapter. For now the important point to understand is that the muscles, tendons, and joints have special detectors called proprioceptors that can tell when these parts of the body are stretching, contracting, and bending. Messages regarding what these body parts are doing are sent to the cerebellum, which does not initiate movement, but monitors and modifies the progress of movements.

The cerebrum calculates what the muscles are doing and determines whether they are acting according to the brain's earlier instructions. The cerebrum then issues new orders to the cerebellum to make any necessary adjustments. The cerebellum then sends the new orders to the muscular system. This process is nearly instantaneous and must be repeated with each muscle movement.

For example, when you try to hit a baseball, first your cerebrum makes the decision to hit the baseball. After processing visual information (from the incoming pitched baseball) and tactile information from the bat you are holding, the cerebrum, through its motor connections, allows you to swing the bat, and the cerebellum modifies your swing in order for you to position the bat in the neighborhood of the ball.

Further complicating your mental calculations as a batter is the pitcher's cerebrum, which is giving instructions to the muscles in the pitcher's arm and hand. The pitcher's cerebellum relays these instructions to the muscles to affect the velocity, spin, and direction of the pitched ball.

In the case of both pitcher and batter, you can see that the cerebral motor cortex and the cerebellum must work closely together in effecting and controlling movements.

By contrast, walking does not require any cerebral involvement. The cerebellum is the key factor in modifying the walking motion. Without the cerebellar influence, you would fall over.

One way we know the importance of the cerebellum in movement is that diseases or injuries affecting it usually cause neuromuscular problems. Many injuries and diseases affecting the cerebellum cause errors in the rate, range, force, and direction of volitional movements and cause coarse, irregular oscillations at the termination of a movement. Basically, people with cerebellar problems lose their balance, suffer from tremors, and become more uncoordinated. Put simply, you'd have a hard time touching your finger to your nose without the "little brain."

Not Really the Middle

The midbrain lies between the cerebrum and the section of the brain stem known as the pons and is about 2.5 centimeters long. It contains many major relay stations as well as many reflex centers. A pair of nuclei called the superior colliculi control reflex actions of the eye, such as blinking, opening and closing the pupil, and focusing. A second pair of nuclei, called the inferior colliculi, control auditory reflexes important in adjusting the ear to sound volume. Nerve fibers originating in the colliculi influence movements of the head and neck in response to visual and auditory stimuli.

A tunnel running through the midbrain and containing cerebrospinal fluid is known as the aqueduct and connects the third ventricle and the fourth ventricle. (Remember these holes in the head from Chapter 5?) The larger part of the midbrain, in front of the aqueduct, forms the cerebellar peduncles.

The cerebellar peduncles are three bands of fibers that connect the cerebellum to the brain stem. The superior peduncle is connected to the midbrain, the middle peduncle to the pons, and the inferior peduncle to the medulla. Information to and from these important structures travel through the peduncles.

At the bottom of the midbrain are reflex and relay centers relating to pain, touch, and temperature. Several regions associated with the control of movements, such as the red nucleus and the *substantia nigra*, are also located in this spot.

> **IQ Points**
>
> The substantia nigra in the midbrain plays a significant role in the causation of Parkinson's disease because of its relationship to the production of the chemical dopamine, a chemical that helps stimulate movement and other reactions.

At the Base of It All

Only 10 percent of your brain stands between you and sudden death. We've talked about the two largest parts of the brain—the cerebrum and cerebellum—and the vital roles they play in the human body, but the starlike structure known as the brain stem ultimately controls the functions that keep us alive, such as breathing, maintaining heart rate, and regulating blood pressure. The brain stem is the area of the brain between the thalamus and spinal cord.

The bulge in the brain stem that looks in drawings of the brain like an Adam's apple is called the pons. The pons is an inch-wide bundle of white matter that forms part of the wall of the fourth ventricle. Consisting of large bundles of nerve fibers that connect the two halves of the cerebellum and also connect each side of the cerebellum with the opposite-side cerebral hemisphere, the pons serves as a bridge (*pons* is Latin for bridge) linking the cerebral cortex and the medulla. Half the cranial nerves, numbers III, IV, V, VI, VII, and VIII, enter the brain in the pons region of the brain stem.

The base of the brain stem is the medulla oblongata, which is at the upper end of the spinal cord. The medulla regulates breathing, heart rate, and blood pressure; it also holds the less glamorous title of vomiting center. The vast majority of sensory fibers coming from various parts of the body cross over to the opposite side of the brain in the spinal cord before they reach the medulla and are relayed to the cerebral cortex, but 80 to 85 percent of motor nerve fibers from the cerebral cortex cross to the opposite side as they pass through the medulla on their way to various parts of the body. Rootlets of cranial nerves IX, X, XI, and XII also emerge from the brain stem in the region of the medulla oblongata.

> **IQ Points**
>
> If you want your kids to impress their friends, teach them that the most important part of the body is the medulla oblongata and let them quiz their classmates on where that part of the body is.

In yet another indication of the difference between the size of a part of the brain and its importance, a piece of brain the size of your little finger filters 99 percent of all the information coming into the brain. This area, which is basically a network of nerve fibers deep within the brain stem, is called the reticular formation.

The reticular formation monitors incoming stimuli and chooses those that should be passed on to the brain and those that are irrelevant and may be ignored. Without this filter, we would be overwhelmed with information and unable to concentrate. This capability to filter is also crucial to our ability to identify danger, which otherwise would be mixed in with all the other incoming signals. In addition to being a filter, the reticular formation controls respiration, cardiovascular function, digestion, awareness levels, and patterns of sleep.

In recent years, the reticular formation has been discovered to be more significant than previously thought. Scientists now believe it to be involved in higher mental processes, in particular the focusing of attention, introspection, and reasoning.

Remembering the Limbic System

The limbic system connects the cortex and the midbrain. The link to the prefrontal cortex is concerned with subjective feelings of emotion, particularly those related to survival, such as sexual desire and the *fight-or-flight reaction*. Researchers have found that stimulating areas of the limbic system can cause patients to feel fear, rage, or excitement. These reactions are not consistent, however, so that entirely different responses may occur when the same region is stimulated again. This inconsistency suggests that no specific part of the brain is responsible for a particular emotion.

The amygdala is one of the basal ganglia and it is located within the temporal lobes on each side of the brain. Composed of two almond-shaped, fingernail-sized structures that have connections to the thalamus, hypothalamus, and pituitary gland, the amygdala helps to filter and interpret incoming sensory information. It also plays a role in initiating appropriate emotional and motivational responses and therefore influences behavior as part of the limbic system. The amygdala is associated with the activation of muscles that produce facial expressions and protective postures, such as crouching. It also can influence the release of adrenaline and other hormones that produce the fight-or-flight impulse. Researchers have found that after the removal of the amygdala a person may not recognize the meaning of expressions of anger such as screams, scowls, and angry voices.

Words of Wisdom

The **fight-or-flight reaction** is the body's natural response to stress in which blood pressure, heart rate, and muscle tension are among the functions adjusted to prepare to confront or evade a threatening situation.

The part of the limbic system that connects to the midbrain also creates a bridge to the reticular formation. The limbic system is also connected to the hypothalamus and other parts of the brain that influence body functions such as heart rate and digestion. This connection helps explain why our emotions can also affect our physiology. For example, when we are anxious or frightened, our blood pressure rises, we sweat, and we can develop ulcers.

The limbic system acts as a thermostat to regulate essential body functions and keep them constant. It is so important and so automatic that it will continue to perform in a comatose person so long as the brain stem continues to function.

The *hippocampus*, located deep within the limbic system, stores and processes memories, helps find information when you want to remember something, and plays a role in emotions. (*Hippocampus* is Latin for seahorse. In classical mythology, the *hippocampus* was a monster with the head and forelegs of a horse and the tail of a dolphin.)

The Glands Have It

Glands produce secretions that can have general or sometimes specific effects on body functions. The secretions are chemicals that carry messages through the bloodstream. These chemical messengers are called hormones. Glands are scattered throughout the body, but the ones we are interested in this book are those in the brain, notably the pituitary gland (also known as the hypophysis).

The Pituitary

The pituitary, which is located in the center of the skull, is about the size of a garbanzo bean. It is attached to the hypothalamus and is responsible for producing a number of important hormones, such as adrenocorticotrophin or ACTH, which plays a role in the body's alarm reaction.

The pituitary was once believed to be the most important gland in the body, the master gland, but the pituitary actually operates under the authority of the hypothalamus, which links the pituitary and the brain. The hypothalamus sometimes responds to the body's needs by nerve impulses and sometimes by directing the pituitary to produce hormones, which are then circulated in the blood to a variety of the body's tissues.

As with so many other parts of the brain, the pituitary gland has two lobes, the anterior and the posterior, each of which releases different hormones that affect bone growth and regulate activity in other glands.

Hormones Secreted by the Anterior Pituitary and Their Effects

Hormone	Major Target Organ(s)	Major Effects
Growth	Liver, adipose tissue	Promotes growth hormone (indirectly), and control of protein, lipid, and carbohydrate metabolism
Thyroid-stimulating	Thyroid gland	Stimulates secretion of thyroid hormones
Adrenocorticotropic	Adrenal gland (cortex)	Stimulates secretion of glucocorticoids
Prolactin	Mammary gland	Milk production
Luteinizing	Ovary and testis	Control of reproductive function
Follicle-stimulating	Ovary and testis	Control of reproductive function

Hormones Secreted by the Posterior Pituitary and Their Effects

Hormone	Major Target Organ(s)	Major Effects
Antidiuretic	Kidney	Conservation of body water
Oxytocin	Ovary and testis	Stimulates milk ejection and uterine contractions

One hormone that the pituitary produces is the growth hormone that affects height. If this hormone is overproduced in a child, it can cause gigantism, or, in an adult, acromegaly (from the Greek words for extremities and enlargement), in which the skeleton gradually enlarges, especially the face, hands, and feet. If too little growth hormone is produced in childhood, it can cause dwarfism.

So what's the difference between someone who is very tall, such as 7 foot, 2 inch basketball player Shaquille O'Neal, and someone suffering from acromegaly? Shaq has slightly higher than normal amounts of growth hormone while someone with acromegaly has an abnormal and persistent increase in growth hormone, which is usually caused by a benign tumor called an adenoma.

Scientists learned to purify growth hormone from the pituitary glands of human cadavers and used it to treat children who were abnormally short. Thanks to modern genetic techniques, it is now possible to create growth hormone in a laboratory. Growth hormone therapy is considered safe, though like all medical treatments it has some health risks. The greater concern has been the ethically questionable use of growth hormone to enhance athletic performance or to try to help children whose stature is within the normal range for their age grow taller.

Gray Matter

The tallest man according to the *Guinness Book of World Records* was Robert Pershing Wallow. He was 8 feet, 11 inches. Wallow was a rather average 8.5 pounds at birth, but by his first birthday weighed 62 pounds. He continued to grow at an extraordinary rate, measuring 6 feet by the time he was eight. He was just under 7 feet tall at age 12. He wore size 37 shoes. Unfortunately, people with acromegaly usually have much shorter than average life spans because the condition causes problems with cardiac function that leads to death. Wallow died at the age of 32 in 1940. His family had almost all of his belongings destroyed so they would not become collectors' items or be displayed as "freak" memorabilia. In case you were wondering, Guinness says the shortest person on record was Gul Mohammed, an adult in India who measured 1 foot, 10.5 inches.

More Than a Pinecone

As you recall, the philosopher Descartes believed the pineal gland was the structure of the mind that controlled the body. This little pinecone-shaped structure at the base of the brain is certainly important, but not in the metaphysical sense Descartes imagined. The pineal gland is involved in monitoring incoming data about lightness and darkness for the regulation of the body's internal clock. Functions associated with cycles, such as sleeping, waking, menstruating, and the onset of puberty, are influenced by the pineal gland. The gland also produces a number of important chemicals, including serotonin, histamine, and norepinephrine. In addition, it is the only source of melatonin, which is believed to play a role in the human body clock. Chapter 11 has more to say about that topic.

The Least You Need to Know

- ◆ The cerebellum is the second largest part of the brain and contains more than 50 percent of all the neurons in the central nervous system. Your ability to maintain your balance and muscle tone is dependent on your cerebellum.
- ◆ The cerebellum is able to automatically coordinate the movements of your muscles and to process information from your eyes and ears.
- ◆ Messages from the muscles are monitored in the cerebellum, which issues orders based on the instructions from the cerebrum.
- ◆ Glands secrete hormones that are critical to promoting growth, controlling reproductive function, conserving water, and handling a number of other essential activities.

Chapter 7

The Real Information Superhighway

In This Chapter

- ◆ Brain wiring
- ◆ Neurons on the head of a pin
- ◆ Giving neurons their space
- ◆ Faster than a speeding impulse
- ◆ Boatloads of messages
- ◆ A chemistry lesson

Think for a moment about how the world communicates through telephone wires and cables that crisscross the land and sea. Without them, we could not talk to each other. Similarly, without a network of "wires" inside our bodies, our brains could not communicate with the other parts of our bodies. This network of wires is called the nervous system.

The body contains 30,000 miles of nerves, which are divided into three types: sensory, motor, and connector. Sensory nerves carry signals to the brain from the ears, eyes, nose, and other sense organs; the motor nerves disseminate messages from the brain to the muscles. Connector nerve cells link the sensory and motor nerves to enable you to make decisions.

Thinking Systematically

The nervous system has three main functions: orientation, coordination, and conceptual thought. These functions overlap and work together.

Orientation relates to the way the body reacts to changes in the environment, both inside and outside the body. In the case of the external environment, stimuli generate signals that are passed along sensory nerves to the brain, which directs a response to the muscles or other body organs. For example, if you are out in the hot sun and touch a metal chair that feels hot, the message "hot!" is transmitted to the brain, which orders the muscles in your arm to move your hand off the chair.

The brain coordinates the process by sorting out the incoming impulses and directing them to the proper places. For example, the brain ensures that a message for the eye does not end up at the elbow.

Conceptual thought relates to the ability of humans to record and store information and draw upon that knowledge as experience on which to base future decisions about responding to changes in the environment. Our brains have the capacity for creativity, abstract reasoning, prediction, imagination, and a host of other capabilities that separate us from the animals.

Orientation, coordination, and conceptual thought are possible because of nerves. Perhaps this is why the system that regulates the body's responses to stimuli is called the nervous system instead of the brain system.

Ancient physicians, particularly Herophilus in the third century B.C.E. discovered that nerves are related to sensation and movement, and the great Greek physician, Galen, identified the nerves as having a role in coordination in the first century, but for many centuries nerves were mistakenly thought to be hollow. The predominant view was that nerves were the route by which pneuma (the spirit), essentially air, traveled through the body.

100 Billion Cells Can't Be Wrong

The basic unit of the nervous system is the nerve cell, also referred to as a *neuron*. About 100 billion neurons are in your brain, and another 100 billion are in the rest of your body. This comparison shows you just how much wiring is concentrated in that head of yours.

Neurons can vary in size and shape, but typically the cell body ranges from 4 to 100 microns wide. (In case you're wondering, 1 micron is equal to one thousandth of a millimeter, so about 30,000 neurons can fit on a pinhead.) The 50 types of neurons that are found in the brain are all slightly different, ranging in appearance from short to long, fat to thin, and so on.

Neurons have two important properties: excitability and conductivity. The former means that they can respond to stimulation. The latter refers to the ability of the neuron to propagate the electrical activity generated by the stimulus. This activity is called a nerve impulse. No other cells in the body can produce nerve impulses.

Gray Matter

Imagining 100 billion neurons isn't easy. Think of it this way: That's about the number of stars in the Milky Way galaxy.

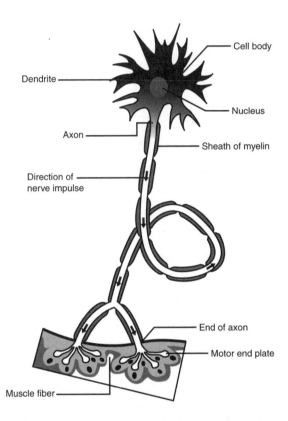

- Cell body
- Dendrite
- Nucleus
- Axon
- Sheath of myelin
- Direction of nerve impulse
- End of axon
- Motor end plate
- Muscle fiber

The neuron has a main cell body with branches called dendrites. The axon extends from the cell body and is the conduit for nerve impulses.

The body (or *soma*) of the neuron is similar to that of all other cells; however, it has some additional unique features. One is a series of small branches or tentacles that extend from the cell body and are referred to as *dendrites* (Greek for tree). A single, long, thin fiber stretching from the body of the neuron is called the axon. In photographs, the neuron looks a little like a drawing of the sun with rays beaming out from its surface and a wire connecting it to the nearest planet. The dendrites receive messages from other neurons and pass them onto the cell body while the axon transmits nerve impulses from the cell body to other neurons. Collections of nerve fibers are called *tracts*.

Words of Wisdom

The word **synapse** was first used in a book on physiology published in 1897 and written by Michael Foster with the assistance of Charles Sherrington. Synapse comes from the Greek *syn* meaning "together" and *haptein* meaning "to clasp."

Like phone lines, some neurons have a kind of insulation that protects them and facilitates their ability to conduct messages. This soft fatty coating is called *myelin*. The neurons that have a myelin sheath are white and comprise the brain's white matter. Myelinated axons also are found in sensory nerve fibers and nerves connected to skeletal muscles. As you may have guessed, the neurons without myelin make up the gray matter.

Motor nerves, sometimes called effector nerves, connect to muscles and glands. They are primarily unmyelinated. Other nerves associated with skeletal muscles are myelinated and are based in the gray matter of the spinal cord and brain.

Unlike telephone wires, the neurons don't actually connect to one another. In Chapter 3, Golgi believed neurons were connected like a spider web, but Cajal later hypothesized that they were separate. Cajal was right: Each neuron is separated by as many as 10,000 tiny spaces called *synapses*. Electrical impulses pass from one neuron to another across these gaps.

We don't have conscious awareness of the process by which the brain transmits the information to our muscles to move because of the Indy car–like speed at which this process occurs. Electrical impulses can travel along neurons at up to 220 miles per hour, or 323 feet per second. The speed varies based on the diameter of the particular axon and whether it has a myelin sheath. The speed in smaller nerve fibers may be a more pedestrian 13 miles per hour.

A typical nerve signal isn't very strong. It is about one tenth of one volt, which is 15 times weaker than a typical 1.5 volt flashlight battery. The magnitude of a nerve impulse coming from a sensory neuron does not determine the strength of a sensation, nor does it determine the strength of a muscle contraction communicated by a motor neuron. Nerve impulses are the same for pain and cold and muscle contractions. The intensity depends on the frequency of the impulse. Generally, the more intense the stimulation, the more nerve impulses are produced. Put another way, if you put your hand on a hot burner, the sensation of excruciating pain won't be produced by one big nerve impulse, but by a large number of rapid ones.

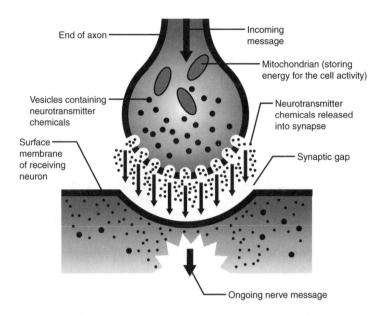

End of axon

Incoming message

Mitochondrian (storing energy for the cell activity)

Vesicles containing neurotransmitter chemicals

Neurotransmitter chemicals released into synapse

Surface membrane of receiving neuron

Synaptic gap

Ongoing nerve message

The synapse is the gap between neurons. Neurotransmitters carry nerve messages across the synapse to other neurons where the signal is converted back to recreate the original message.

Closing the Gap

Like a telephone system, the messages in the body are carried by electrical signals. These signals fly down the length of the axon and then, like a car racing down a road that runs smack into a body of water without a bridge, must find a way to cross the synapse. The impulse can't be transmitted electrically; first, it needs to be converted to a chemical form.

We could wow you here (or put you to sleep) with a complicated explanation of ions, voltage, sodium channels, potassium, and the other elements that go into explaining the electrical signals neurons use to communicate, but we'll leave that to the authors of *The Complete Idiot's Guide to Chemistry and Physics.*

For our purposes, what's important to understand is that the body has "boats" that carry the electrical messages to other neurons. The electrical impulse reaches the neuron's boathouse, a balloon-like structure called a vesicle, which releases a naturally occurring chemical called a *neurotransmitter*. This neurotransmitter is the boat that ferries the signal across the synaptic cleft (the space between synapses) in thousandths of a second.

The dendrite of the nearby neuron has a dock, or receptor, to receive the boat, and each boat has a key to a particular dock. At this point the neurotransmitter combines with other chemicals at the receiving neuron to produce a new electrical signal.

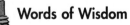

Words of Wisdom

A **neurotransmitter** is a substance that transmits nerve impulses across a synapse.

A Whole Lotta Synapses

A single neuron can have as many as 100,000 synapses constantly sending and receiving messages from hundreds of thousands of neurotransmitters released by 100,000 surrounding neurons. To put this number in perspective, consider that the Amazon rain forest contains approximately 100 billion trees, which is roughly the number of neurons in the brain. To imagine the number of connections between the neurons, think of the leaves on all of those trees. Someone figured out that the number of different possible connections would be 40 quadrillion, or a 4 with 16 zeroes.

Researchers have found more than 50 types of neurotransmitters, Among the most common are acetylcholine, dopamine, serotonin, and norepinephrine. Particular neurotransmitters are located in specific areas of the brain. For example, the neurons that release dopamine are produced in the midbrain.

Neurotransmitters are divided into two main classes: exciters and inhibitors. The former trigger the receiving neuron to accept a signal while the latter blocks it from being received. These chemicals occur naturally in the body, but they can also be introduced with drugs. For example, caffeine acts as an exciter by making it easier for impulses to cross a synapse. Other drugs, such as morphine, can inhibit a transmission to help prevent or relieve pain.

Remember that the neurotransmitters have matching receptors. These receptors are now known to belong to families, so a group of receptors of a particular family are specific to one neurotransmitter. By learning about these relationships, researchers may be able to develop drugs targeted for specific receptors to treat various disorders.

The balance of these chemicals influences the functioning of the brain. Several neurotransmitters have been found to play a role in specific behaviors. For example, a lack of serotonin is believed to influence aggression. Endorphins are involved in producing sensations of pleasure. Schizophrenics seem to have a lack of dopamine, and Parkinson's disease, too, is characterized by a significant decrease in dopamine.

The relationships between the chemical and electrical properties of the brain and behavior are the basis for the field of physiological psychology.

Fire When Ready!

When a neuron is stimulated, it acts according to an all-or-nothing principle by which it either does not transmit an electrochemical message or it fires one at maximum strength down the length of the axon. Notice we said electrochemical message. That means it's time for a quick chemistry lesson.

The chemicals in your body are called ions because the atoms that comprise them have electrical charges. The important positively charged ions are sodium, potassium, and calcium. Other ions, such as chlorine, have negative charges.

With us so far?

The natural tendency is for the ions inside and outside the nerve cell to balance each other, but the neuron has a membrane surrounding it that allows certain ions to pass through, but blocks others, which may create an imbalance. When a neuron is not sending a signal, it is at rest with relatively more sodium and potassium ions outside the neuron and more chlorine ions inside the neuron creating a negative charge. The difference in voltage between the inside and outside is called the resting potential. The resting potential is -70 millivolts (mV), which means that the inside of the neuron is 70 mV less than the outside.

When a nerve is stimulated, the permeability of the cell membrane changes quickly, producing what is called depolarization. An exchange of ions then takes place across the membrane. Sodium ions (the ones with the positive charge) rush into the neuron, which makes the cell become more positive. As this change occurs, the resting potential rises toward 0 mV. When the depolarization nears -55 mV, an impulse called an action potential fires. The impulse fires only if the action potential is strong enough to reach this threshold; if not, the stimulus was a dud, and the signal is not transmitted.

Soon potassium begins to gush from the neuron and reverse the depolarization. This process is referred to as the refractory period when the cell returns to its resting potential.

Whew! Following that explanation is more fun than humans are supposed to have. Don't worry, there's no test at the end of the book.

The Forgotten Brain Cell

Neurons get all the publicity, but they are outnumbered by as much as 50 to 1 by another type of cell known as the neuroglia (literally "nerve glue"), handily nicknamed glia. These cells perform a variety of functions for the neuron, including helping them recover from injury, make synapses, and produce myelin. They also physically support the neurons.

There are several types of glia, and each type performs a different function. For example, Schwann cells and Oligodendroglia provide the myelin to neurons in the peripheral and central nervous systems respectively. The star-shaped Astrocytes provide nutrients to neurons, keep them in place, digest parts of dead neurons, and handle a number of other support functions for the neurons. Glia do not have axons or synapses.

Snakebitten

What makes poisonous animals such as snakes, spiders, and scorpions dangerous to humans? The answer is that the stings and bites from these creatures, which are designed to protect the animal from predators or to kill prey, release neurotoxins that have serious effects on our nervous systems. The neurotoxins either block or activate the release of chemicals from neurons and can cause paralysis, suffocation, and heart failure. Though it is rare, people do sometimes die from the effects of neurotoxins. Some of these poisons have killed people in as little as five minutes. About 8 to 15 people in the United States die from snakebites each year.

There are about 250 venomous sea creatures and 300 venomous snakes (out of 3,000 species). Some of the more dangerous animals include the stonefish (the most poisonous fish), the blue-ringed octopus, the pufferfish, the black widow spider (the fearsome looking tarantula is not deadly), the poison arrow frog, the scorpion, and a variety of snakes including the Mojave rattlesnake, black mamba, and cobra. The most dangerous snake is either the taipan, hook-nosed sea snake, or Russel's viper. The spitting cobra can, as the name implies, spit its venomous neurotoxin at an enemy and hit it from as far as 10 feet away.

In the movies, whenever someone is bitten by a snake, the person trying to be helpful tries to suck out the venom. In reality, you should never do this because the venom will already be in the victim's bloodstream by the time you start sucking. Instead, the wound should be wrapped with a pressure bandage (not a tourniquet). It won't be easy, but the victim should also be kept calm. Most important, get to the hospital immediately!

Now for the Hard Part

As you can see, we've learned a great deal about the functioning of our nerves and the way the brain receives and transmits information through the body. The analogy in this chapter of the telephone network raises an interesting question: Does the brain have connections similar to a cellular phone network? Perhaps in the future we will find out that many of the connections in the brain are unknown because they are wireless.

Truth be told, we still don't know beans. No one knows how all the transmissions between the billions of neural connections create memories, allow us to reason, or produce emotions. Brain research is proof of the axiom: The more you know, the more you know you don't know.

Now that you have an understanding of the components of the brain and an idea how they all work, it's time to explore human behavior and find out a little more of what we do know about the relationship between the brain and the ability to think, feel, speak, hear, see, taste, smell, move, and work.

The Least You Need to Know

- The brain contains 100 billion nerve cells known as neurons.
- Neurons are not connected; they are separated by tiny gaps known as synapses.
- Nerve impulses must be converted from electrical to chemical impulses to cross the synapse.
- Neurotransmitters are released by the neurons to carry signals from one neuron to another, where other chemicals convert them back to electrical impulses.

Chapter 8

You've Got Nerve(s)

In This Chapter

- ◆ Easy as I, II, III
- ◆ Seeing the big "E"
- ◆ More than a pretty face
- ◆ Balance you can hear
- ◆ Body controls

When you smile, frown, chew, taste, see, and hear, a dozen special nerves are involved in relaying sensations to the brain and communicating messages to the muscles from the brain. These nerves are the cranial nerves. The twelve pairs of cranial nerves emerge from the base of the brain; all but the first two pairs are attached to the brain stem.

These nerves connect the brain to the parts of the head and neck, such as the ears, eyes, nose, mouth, and face. This chapter explains which parts of the body each pair of cranial nerves affects.

What's Your Function?

Some nerves control muscle movements, others convey information from the sense organs, and others combine the two functions.

The motor components of the cranial nerves send their axons out of the cranium to control general muscle movements, (such as eye and facial movements) and specialized muscles (such as the heart). The sensory components come from outside the brain, specifically from collections of cells called sensory ganglia. These ganglia are divided into branches: one branch is connected to a sensory organ, such as the taste receptors in the tongue, and transmits signals through the ganglia to a second branch that penetrates the brain. The following table lists each cranial nerve pair and its function.

Summary of Cranial Nerve Functions

Nerve	Function
I. Olfactory	Handles smell
II. Optic	Handles vision
III. Oculomotor	Controls several muscles that move the eye and affect the size of the pupil
IV. Trochlear	Controls a muscle that moves the eye
V. Trigeminal	Carries messages of sensation from the face, nose, and mouth, mediates the corneal reflex, and carries motor fibers to the muscles of mastication
VI. Abducens	Controls another muscle that moves the eye
VII. Facial	Controls muscles in the face, ear, the pharynx, and four of six salivary glands and carries some sensation from the tongue
VIII. Vestibulocochlear (Auditory)	Deals with balance and hearing
IX. Glossopharyngeal	Carries sensation from part of the tongue and controls some muscles involved in swallowing and the pharynx
X. Vagus	Controls heartbeat, breathing muscles, and some liver and kidney functions
XI. Accessory	Controls some neck muscles
XII. Hypoglossal	Controls tongue muscles

Numero Uno

When you stop and smell the roses or wake up and smell the coffee or get attacked by one of those perfume spritzers in a department store, cranial nerve I goes into action. The cranial nerve I, (cranial nerves are normally identified by a Roman numeral) also known as the olfactory nerve, is comprised of about 20 to 24 neurons and is connected at the uppermost part of the base of the brain to the smelling cells in your nose. If this nerve is damaged by injury or disease the result can be anosmia, the loss of the ability to smell. This loss can also dull the sense of taste.

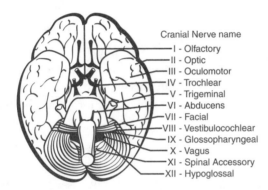

Cranial Nerve name
I - Olfactory
II - Optic
III - Oculomotor
IV - Trochlear
V - Trigeminal
VI - Abducens
VII - Facial
VIII - Vestibulocochlear
IX - Glossopharyngeal
X - Vagus
XI - Spinal Accessory
XII - Hypoglossal

The twelve pairs of cranial nerves.

For Your Eyes Only

Cranial nerves II, III, IV, and VI are all related to the ability to see. The second cranial nerve is the optic nerve. The optic nerve is about 50 millimeters long and contains more than one million fibers, which is significantly more than any of the other cranial nerves. It connects directly to your eyes and transmits visual information to the brain.

The optic nerve from each eye meets at a junction called the *optic chiasm*. Here, the fibers from the nasal halves of each retina cross over to the opposite hemisphere while the fibers from the temporal (toward your ear) halves of the retina project to the same side of the brain. The fibers that extend from the chiasm to the lateral geniculate body in the thalamus are called the *optic tracts*, and they contain information from both eyes. If an optic tract on one side is damaged, it can produce partial blindness in both eyes. The part of the system that projects fibers from the thalamus to the cortex is called the *optic radiation*.

Looking Reflexively

When you think of reflexes, the first thing that comes to mind is probably the doctor banging your knee with a hammer to see whether or not it moves. It may never have occurred to you that your eyes also have similar reactions called optic reflexes. The four optic reflexes are the light reflex, the visual fixation reflex, the protective reflex, and the near point reaction reflexes.

The light reflex involves the optic nerve and others in constricting the pupil in reaction to changes in the amount of light entering the eye.

The visual fixation reflex not only involves the optic nerve, but also cranial nerves III, IV, and VI. This reflex adjusts the eyes to bring images into correct alignment when you are performing tasks such as following the lines in a book when reading.

What happens when something is poked in the direction of your eye? Usually, you immediately blink. If the stimulus is strong enough, you might not only blink, but also reflexively raise your hands in front of your face for additional protection. Thus this reflex is called the protective reflex.

Near point reaction reflexes come into play when you try to focus on a close object. A series of actions are required, including thickening the lenses of the eye, the convergence of the eyeballs to bring the object into alignment on the retina in each eye, and the narrowing of the pupils to allow the appropriate amount of light into the eye in order for the object to be seen clearly.

When the Camera Breaks

Injury to an optic nerve can cause total blindness in that eye. A number of tests are used to check for the proper functioning of the optic nerve. The two most familiar are the Snellen eye chart and the Ishihara plates. Don't recognize the names? That's okay; you've taken both tests. The Snellen test is the chart with the big "E" at the top that tests for visual acuity. The Ishihara plates are a series of pictures with circles full of different colored dots that test for color blindness. If you see one number in the pattern, your vision is normal; seeing a different number indicates color blindness.

Rolling Your Eyes

The oculomotor nerve (III) controls the muscles that turn the eye up, down, and inward, cause the pupil to constrict, and raise the upper eyelid. The trochlear nerve (IV), the smallest cranial nerve, interacts with the muscle that turns the eye down and out. The abducens nerve (VI) also connects to a muscle that rotates the eye outward.

An injury or a lesion of any of the cranial nerves (III, IV, and VI) that control eye movements can cause double vision. Damage to the oculomotor nerve can have a number of effects on the eyes. One is to cause the eye to rotate downward and slightly outward; another is to cause drooping eyelids. In particular, diabetes may cause an abnormality of the oculomotor nerve that causes the eye to move outward and the pupil to dilate. Damage to the trochlear nerve can cause a loss of the ability to normally rotate the eye. Damage to the abducens can also cause the attached eye to become crossed (as in being cross-eyed).

IQ Points
When police officers stop someone and administer a roadside test for inebriation, they check a driver's eye movements for signs of nystagmus (a rapid involuntary oscillation of both eyes when looking far right or far left). This condition is usually found in association with ataxia, which is difficulty walking a straight line. These may be signs of intoxication.

One way to test how well the oculomotor, trochlear, and abducens nerves are functioning is to ask someone to stand in front of you and hold his head still. Hold your finger in front of his face and ask him to follow it with his eyes, up and down, right and left. If the person doesn't have any trouble following your finger, the nerves are working normally. (Does this seem like a test you may have already had at a doctor's office?)

A Sensitive Face

How do you smile, frown, purse your lips, and knit your eyebrows? You make these movements with the help of cranial nerve VII. For some time, scientists believed that the same cranial nerve also controlled the muscles involved in chewing, but Charles Bell demonstrated early in the nineteenth century that the sensory and motor functions of the nerves were separate and that the trigeminal nerve (V) is the one responsible for mastication.

The trigeminal nerve, the largest of the cranial nerves, has three branches and connects to the sensory receptors in areas of the face, such as the forehead, jaws, and sinuses. The ophthalmic nerve branch carries sensory information from parts of the eye, the brow, and inside the nose. It also carries fibers responsible for the corneal reflex (eye blinking in response to the cornea being touched). This nerve branch is divided into still smaller branches with even more specific functions. For example, the frontal nerve is attached to the skin on the upper eyelid and the forehead. The maxillary nerve branch is also very specific, conveying sensation from facial areas such as the lower eyelid, side of the nose, upper lip, upper jaw, and teeth. Finally, the mandibular branch of the trigeminal nerve interacts with the muscles of mastication and enables you to chew your food. This branch is also involved with sensation in the lower jaw, teeth, the cheek, and two thirds of the tongue.

Code Blue _____

Bell's palsy is a facial nerve disorder that causes a partial paralysis on one side of the face, usually affecting the eyelid, forehead, and the muscles moving the lips. The cause of this disorder is unknown, but scientists suspect that a reaction to a virus causes the facial nerve within the ear bone to swell. The disorder affects an estimated 40,000 Americans each year, and although about 80 percent of the people with Bell's palsy recover completely within three months with or without medical treatment, some sufferers never recover and have no assurance of a cure with treatment. In some instances, high doses of steroids have been successful in restoring normal facial nerve function. In some rare cases, removal of part of the bone of the facial canal through which the facial nerve runs has restored facial nerve function before complete paralysis has occurred.

Why can't more parts of the body have simple descriptive names like the facial nerve (VII)? This nerve specifically controls facial expression and also plays a role in regulating the movements required in speech and chewing. If you look in the mirror and can make funny faces, smile, and frown, your facial nerve is working properly.

Sensations from about two thirds of the tongue are transmitted to the brain via the facial nerve, which is also involved in controlling four of six salivary glands and the muscles of the pharynx. A simple test for the sensory component of the facial nerve is to put something sweet or salty on the tip of your tongue to check whether you can taste it. The facial nerve also controls the smallest muscle in the body, the stapedius, which is inside the ear.

Testing 1, 2, 3 ...

How can you have one cranial nerve called the facial nerve and then name the next one vestibulocochlear? This name is a mouthful, but if you break the word into its components, it does give a clue about the nerve's function. As you may recall from the previous chapter, the vestibular system has to do with coordination and balance, so the vestibular component of cranial nerve VIII conveys impulses to the brain related to equilibrium and position, as well as movements of the head and neck. The *cochlear* or auditory part of the nerve is involved in hearing.

Words of Wisdom _____

From the Latin for "snail," the **cochlea** is a part of the inner ear that is coiled like a snail shell and is responsible for hearing.

To get a better sense of what is called the vestibular reflex, consider what happens when you are sitting in a car staring out the window. The scene moves quickly past you, but you can still see it for some time because your eyes adjust by moving in the opposite direction. Think about the test we previously suggested in which someone holds up a finger in front of you and moves it to see whether your eyes can follow it. Try a modification of that test and have the person keep his finger still and move your head from side to side and up and down instead. Your eyes will move in the opposite direction of the finger. This reflex is similar to a brain stem reflex known as *doll's eye movement*; that is, a movement of the eyes opposite from the side to which the head is moved.

Damage to the vestibular part of the nerve can cause eye movement disorders, nausea, and vertigo. If the cochlear nerve is injured, tinnitus (ringing in the ear) and/or deafness can occur.

> **Code Blue**
>
> Motion sickness may be due to a prolonged and excessive stimulation of the vestibular system. Incidentally, no one knows why some people get motion sickness and others don't. Even those who are sure they can never get sick can be made sick with the proper stimulation (just ask any astronaut).

Gulp and Inhale

Several cranial nerves combine to control certain functions. For example, cranial nerves IX, X, and XI are involved in regulating the muscles for swallowing and moving the diaphragm for breathing. These three cranial nerves have other duties as well, as explained in the following sections.

The Ear, Tongue, and Throat Nerve

The glossopharyngeal nerve (IX), which emanates from the medulla just above the vagus nerve (X), controls some swallowing muscles and is involved with the parotid salivary glands. It also plays a role in transmitting taste sensation from the rear third of the tongue and other sensations from the back of the mouth and pharynx.

You can check the sensory component of cranial nerve IX using the same test suggested for the facial nerve, only this time put something sweet or salty on the back of your tongue and see whether you can taste it.

Nerve IX also is responsible for pain and temperature from the ear canal and the skin of the ear. It also has a branch to the carotid artery in the middle of the neck that ends on a special organ called the carotid sinus, from which it carries a pressure message to the medulla for the control of blood pressure. And if you sing the word glossopharyngeal to a bossa nova beat, it sounds quite catchy.

Viva Las Vagus

The *vagus* nerve lives up to its name by supplying neural connections to a wide variety of body organs, from the intestines to the eardrum. The old X nerve sends branches to the bronchial tree (the tubes that carry air to various parts of the lungs), inhibitory fibers to the heart to slow the heart rate, and motor fibers to the larynx, esophagus, stomach, small intestine, and bile duct. It also sends secretory fibers to the stomach and pancreas.

Words of Wisdom

Vagus is Latin for "wandering." The longest cranial nerve has this name because it wanders down the body.

Stimulation of the vagus nerve causes a number of events, including the following:

◆ Decrease in the heart rate

◆ Constriction of the bronchial smooth muscle (partly responsible for breathing)

◆ Stimulation of the glands of the bronchial mucosa (the lining of the bronchial tubes that becomes inflamed with bronchitis and constricted with asthma)

◆ Promotion of peristalsis (the process that makes food move along) in the gastrointestinal tract

◆ Relaxation of the pyloric (the muscle between the stomach and the small intestine) and ileocolic (the muscle between the small and large intestines) sphincters

◆ Stimulation of the secretion of gastric and pancreatic juices

Sensory branches of the vagus nerve conduct sensory stimuli from the heart, bronchi, esophagus, stomach, small intestine, and ascending colon. Last, but not least, vagus stimulation may cause nausea. That's right, it's the vomit nerve. (Sometimes even *Idiot's Guides* tell you more than you want to know.)

Damage to one vagus nerve can cause difficulty in swallowing and speaking. If both vagus nerves are injured, you're in a heap a trouble because this condition can cause life-threatening paralysis of the muscles in your throat that can lead to the inability to breathe.

IQ Points

While snacking on a pretzel watching a football game, President George W. Bush fainted. He later joked that he should have listened to his mother when she said to chew before swallowing, but the incident itself wasn't funny. How could someone be rendered unconscious by a pretzel? It's not as unusual as you might think; there's even a term for what happened to Bush: neurally mediated vasovagal syncope. In layman's terms, the president's heart rate fell because the pretzel didn't go down right. More technically, the vagus nerve was stimulated, and this stimulation caused the decrease in his heart rate. This same problem can occur if the nerve is stimulated by fear, intestinal cramps, or unpleasant smells or sights, such as the sight of blood (which is why people sometimes pass out watching a procedure in the hospital or getting a shot).

The Right Accessory

Given all the vital parts of the brain, you might think that the nerve given the unglamourous name of accessory (XI) can't be too important. Judge for yourself: This cranial nerve controls the neck muscles.

Stick Out Your Tongue

The hypoglossal nerve (XII) controls the tongue muscles. To check whether yours is working stick out your tongue and move it around. Hint: you might want to conduct this test alone.

The Least You Need to Know

- ◆ Twelve pairs of cranial nerves connect the brain to critical parts of the body to receive sensory information and send instructions.
- ◆ Four cranial nerves help you to see. The optic nerve (II) is most directly involved in the sense of sight; the oculomotor (III), trochlear (IV), and abducens (VI) nerves are primarily responsible for eye movements.
- ◆ The trigeminal nerve (V) sends sensory information from your face to the brain while the facial nerve (VII) controls facial movements, such as smiles.
- ◆ Swallowing and breathing require the assistance of three cranial nerves: the glossopharyngeal nerve (IX), the vagus nerve (X), and the accessory nerve (XI). These nerves have other duties as well.

Part 3

The Human Computer

We trust you survived the rigors of our anatomy lesson and feel prepared to learn about the way all those parts of the brain work together to help us live our lives. This part of the book explores the process of thinking, the ability to speak, the senses, voluntary and involuntary reactions, and basic human needs. It's really quite exciting.

Chapter 9

Speak Now!

In This Chapter

- ◆ The origin of speech
- ◆ Baby talk
- ◆ Speaking in tongues
- ◆ Think first
- ◆ See Spot run

What separates humans from animals? One trait is our ability to communicate our thoughts, feelings, and ideas through the use of a spoken language. Our ability to learn words, put them together into ideas, and store verbal information is all rooted in our brains.

How cool is that?

That phrase itself dates the person who uses it and illustrates how language can stay in our memories. (Memory will be discussed later, if we don't forget.)

Most of what we know about how the brain functions is a result of animal experiments. Because speech is a human activity, however, animal research offers little or no insight in this area. Experiments by Penfield and others (which are described in Chapter 4) in which a scientist stimulated the brain of a conscious patient to test the effect on speech have been crucial to our understanding of the areas of the brain involved in language and recognition. This chapter provides an overview of what we know about how we talk.

The Origin of Babble

No one knows the origin of speech, though there is naturally a great deal of speculation on the topic. The Bible says that the first man, Adam, gave names to all the animals, and according to the Book of Genesis, the "whole earth was of one language, and of one speech."

That all changed with the Tower of Babel. Again, according to the Bible, God became angry when an effort was made to build a tower that would reach the heavens. God feared that people would begin to believe they were godlike and decided to make sure they could never complete their project by making them speak different languages so they could not communicate with each other. The resulting confusion led the people to give up on the tower and scatter around the world to build cities where the various languages would be spoken.

Many people do not accept the idea of a divine origin of language. Lots of other theories have been proposed. We particularly like Max Mueller's (1823–1900) description of four theories that he referred to as the ding-dong, bow-wow, pooh-pooh, and yo-heave-ho theories. The ding-dong theory originated with the philosopher Plato and makes a connection between the words and the things they stand for. The bow-wow theory holds that words are based on sounds that imitate their objects. The pooh-pooh theory suggests that speech was initially involuntary sounds produced when primitive humans became emotional. Finally, the yo-heave-ho theory maintains a connection between sounds and communal effort.

Gray Matter

Though we typically recognize words as single sounds, the brain processes words as several individual sounds. These individual sounds, the building blocks of language, are called phonemes, and they can be combined to create thousands of words. In English, the 26 letters of the alphabet singly or in combination form 44 phonemes.

Whatever the origin of speech, the ability to communicate is a complex process that involves multiple areas of the brain. Adults often take the ability to speak for granted, but the difficulty of the task becomes apparent when you have children and see first-hand the stages of communication, from "goo-goo" to "mama" to "I want" to "Pseudoantidisestablishmentarianism."

Another way to get an idea of the complexity of speech is to listen to a synthesizer or other man-made device that produces speech. Though the quality has vastly improved over the years, most of these machines sound more like rusty tin cans than the seemingly human HAL in *2001: A Space Odyssey*.

From the Mouths of Babes

Though parents often panic if little Jane or John doesn't begin to recite Shakespeare at age one, most children do learn how to speak, and they learn at the time that is right for them. Most language activities have no set developmental timetables, though all the parenting books lead new mothers and fathers to sometimes worry unnecessarily if their children are not where they are "supposed to be."

Typically, children go through a series of stages in language development. The first sound babies make is what is characterized as an undifferentiated cry, usually a reflexive response to some stimulus. When babies make this cry, frustrated parents turn to each other and ask, "Why is the baby crying?" They have no way of knowing whether the baby is hungry, constipated, hurt, or tired. Just imagine how frustrated the baby must be. He's probably thinking, "Don't those big people know that I'm freezing my butt off and I need my blanket wrapped tighter?"

After the first month or so, the baby still cries, but now the parents can sometimes figure out the reason. They may sense that the baby is hungry, and when they respond to the cry with food, voilà, the wailing stops. The baby gradually moves on from crying to experiment with other sounds.

Bibble, Bibble

Toward the end of the second month, infants begin to babble, and apparently believing this babble to be a form of language babies must understand, the parents start babbling back.

People who study *linguistics* have found that babbling is an important learning process, which has been likened to the tuning of a musical instrument. The baby hears the sounds it is making and associates them with the physical movements required to make them. Interestingly, children who are born deaf begin to babble, but soon stop because they don't hear themselves.

Roughly in the ninth or tenth month, children begin to imitate the sounds that they hear others make. This mimicry is the prerequisite for speech and is a more complex process than it appears to be. Think about a mother and father saying something as simple as "Dada" to a baby. It will sound different to the infant when the father says it than when the mother says it. And it will sound different when the child mimics the word. It's remarkable that a baby can learn to recognize and replicate a pattern of sounds at all.

Parents also unconsciously speak differently to their babies than they ordinarily do. Yes, the baby talk part is conscious, but the tone is not. As the PBS series "The Secret Life of the Brain" explained, parents speak in "caretakerese," a tone that is about an octave higher in pitch and slower than our normal pattern. The particular signals we transmit are apparently what babies need to develop the neuron connections required for speech.

Words of Wisdom

Linguistics is the scientific study of language. One branch of this field, neurolinguistics, specifically examines how language is processed and represented in the brain.

In the first 12 to 18 months, infants begin to learn individual words and phrases; they learn more descriptive words in the six months before their second birthday. From age two on, children begin to speak in sentences, and before you know it, they're jabbering nonstop, and you begin to have occasional thoughts about the good old days before they learned to speak. If a child isn't speaking by the time he or she turns three, there could be a problem, and physician should check for physical reasons for the delay, but sometimes the child is just developing more slowly.

"C" Is for Cat

Another important but unexplained issue in speech development is the recognition of words. A single word such as *cat* can be pronounced in different ways, and it will sound different when spoken, for example, by someone with an accent or a high-pitched voice. Yet somehow a child learns to recognize the word as *cat* and relate it to an image of a feline or a live animal.

This process of "naming" becomes even more complex as time goes on. It starts with associating a word with an object, but later, the relationship between words and more abstract concepts must be learned. For example, the word *cat* no longer is just a picture in a book of a cat, but an animal that has certain characteristics, such as fur and whiskers, and makes a purring sound.

The brain must also interpret the context of language. For example, if you were asked to spell the word *write*, you would have to know the question referred to the process of putting pen to paper or you might spell it *right*. Similarly, you could not tell what the word *right* meant if someone started to say, "The right ..." until you heard more of the sentence. If the sentence continued as, "The right answer to the question ..." you would know the meaning of *right* was "correct." If, however, the sentence began, "The right side of the brain is known as ..." the meaning would be "the opposite of left." We wish we could tell you how the brain manages to figure out which right is right, but no one has figured it out yet.

It's All Greek to Me

We are capable of speech because our bodies have the structures, such as our vocal cords, necessary to produce sounds; and because our brains can comprehend the grammatical principles on which language is based. What is especially remarkable is the fact that children are not wired to speak a particular language; they can learn whatever language is spoken around them.

IQ Points

One study in which brain scans were done on children ages 3 to 15 showed that different parts of the brain grow at different rates. The researchers had expected to find that all areas of the brain grew at the same rate. One possible implication for the finding was that the best time to learn a second language was from 6 to 13 because this period is a time of rapid development of the parts of the brain dealing with language.

The physical process of speaking involves the coordination of the muscles of articulation. The nerves from the lower part of the left frontal lobe cross through the corpus callosum into the right hemisphere. Signals also pass through the pons and medulla and are transmitted to the cranial nerves that control the lips, tongue, soft palate, and larynx. Other nerves stimulate the muscles that move the diaphragm.

Sounds are made when air is exhaled from the lungs as it passes between the vocal cords. Vibrations of the vocal cords produce sounds that are amplified by the voice box. The tongue, teeth, palate, and lips shape the passing air to produce various sounds.

In the course of learning their native language, children adopt the pronunciation of their teachers. For example, children can say the word rice and properly pronounce the "r." Japanese children, however, learn to speak in a way that does not distinguish an "r" from an "l," so they grow up pronouncing "rice" as "lice" and may never be able to overcome this experience even if they later learn English.

If you want to start mastering other languages, you might be interested to know that there are more than 5,000 to choose from. If you want to speak to the most people, pick Mandarin Chinese, which is spoken by approximately 850 million people, more than double the number who speak English. (Incidentally, a man from New Zealand mastered 58 languages.)

Back to Broca and Wernicke

To review for a moment what we learned earlier, the language center of the brain, known as Broca's area, is in the left hemisphere in right-handed people and vice versa. Well, that's not exactly true after all. Actually, language is under the control of the left hemisphere in 97 percent of all people. Children, however, appear to use both hemispheres during their early language development before one side, usually the left, becomes dominant.

Sometimes, a person whose left hemisphere is damaged can still speak because the opposite hemisphere adopts that function, but this adaptation becomes less common after about age 10. More often, if Broca's area is damaged, a person will lose the ability to speak or write. Damage to Wernicke's area of the brain leads to a condition whereby the ability to understand written or spoken language is lost.

Broca's area is also involved in controlling parts of the body involved in enabling us to speak through various intermediary centers in the thalamus through which messages pass to and from the cranial nerves involved in the formation of sounds and words. The cranial nerves also control the muscles of the mouth, pharynx, and larynx. The nerve fibers connecting the ears and brain are also critical as they pass along sound-generated impulses to the temporal lobe of the brain where they are analyzed, remembered, and translated into recognizable language. These signals also allow the brain to control the volume, pitch, and content of speech.

> **IQ Points**
>
> Why is language controlled primarily by one hemisphere of the brain? A variety of ideas have been put forward, such as the notion that this structure is more efficient because information needs to be transferred over shorter distances with fewer connections. Another suggestion is that this structure is more efficient because it differentiates the hemisphere of the brain that processes language from the one that controls the muscles involved in speech. These hypotheses are interesting, but still unproved, which leaves us with yet another unsolved mystery.

The ABCs

As you read this book you are not even aware of the vast neural network that is firing in your brain as the cerebellum directs the movement of your eyes and messages from the eyes are traveling along the optic nerve to the cortex where the information on the page in front of you is being interpreted. The occipital area is processing all the visual information, such as the words, letters, and pictures, while the frontal lobe is searching your memory for what these images mean and how they relate to your existing base of knowledge. The speech area of your brain is active as well, even though you are not speaking.

> **Gray Matter**
>
> Researchers have found that third grade is the critical point for mastering basic reading skills. Three out of four children who are still having difficulty reading at that point will continue to be poor readers when they become adults.

Researchers have found that the ability to read is related to the capacity to understand spoken language and to translate text on a page to speech. Most children learn to read regardless of the method used to teach them, but about 10 million children have difficulties learning to read, and this difficulty can have lifelong consequences, from a loss of self-esteem to difficulty reading a map or following safety instructions.

Help Needed

Reading is a function of the communication between the visual centers of the brain and the language areas. When brain damage or a lesion somehow disconnects these areas, a language disorder can result.

An estimated 15 to 20 percent of the population has a reading disability. Several reading disorders have been identified; the following four definitions come from the International Dyslexia Association:

- ◆ **dyslexia**   A language-based disability in which a person has trouble understanding words, sentences, or paragraphs; both oral and written language are affected.

- ◆ **dyscalculia**   A mathematical disability in which a person has unusual difficulty solving arithmetic problems and grasping math concepts.

- ◆ **dysgraphia**   A neurologically based writing disability in which a person finds it hard to form letters or write within a defined space.

- ◆ **auditory/visual processing disorders**   Disorders in which a person has difficulty understanding language despite normal hearing and vision.

Dyslexia is probably the best known of these disorders and is often associated with a person's tendency to mix up letters or see words as though they were written backwards. (Actually, most beginning writers have problems with reversing letters.)

Code Blue

Attention Deficit Disorder (ADD) and Attention Deficit Hyperactive Disorder (ADHD) are behavioral disorders, not learning disabilities.

Say What You Hear

When children lose their hearing, it severely hinders their ability to speak. They tend to make sounds that duplicate what they hear, which may come through to them as distorted or muffled. If the neural structures for hearing are intact, a person can hear their own voice through the bones of the skull, but not the sounds of others. This form of hearing loss often leads the affected person to speak softly. The opposite is true of those with neural damage, who cannot hear themselves any better than those around them and thus often overcompensate by speaking loudly.

People who are totally deaf usually cannot learn to speak on their own. Special schools and therapists can teach a deaf person to speak by showing them how to make the proper sounds. This practice is controversial in the deaf community where some activists maintain that sign language is an equally legitimate form of communication.

Perhaps as much as 5 percent of the U.S. population suffers from some form of speech disorder. These people fall into a variety of categories, including problems with pitch and volume, development of language skills, and stuttering and stammering. The causes of these problems include brain injury, disease, and psychological problems.

Diseases such as Parkinson's and cerebral palsy affect the brain's ability to control the muscles required for speech. Consequently, sufferers of these conditions often have difficulty speaking clearly; their voices may tremble or be difficult to understand.

One speech disorder related specifically to the brain is aphasia, a condition whereby a person has difficulty expressing thoughts and understanding the language of others. This condition is a result of brain damage. To test for aphasia, patients are asked to perform a number of ordinarily simple tasks:

◆ Speak fluently with a normal rhythm and without grammatical errors.

◆ Accurately repeat spoken sounds, words, and phrases.

◆ Understand spoken language and follow spoken commands.

◆ Consistently name common objects that are presented visually, verbally, or tactilely.

◆ Read aloud accurately with comprehension.

◆ Name words spelled aloud.

◆ Write legibly and grammatically.

Depending on the overall result of these tests, a physician can often determine whether a patient is suffering from aphasia and which part of the brain may be damaged. For example, if the patient speaks in a normal, fluent pattern, uses incorrect or nonexistent words, abnormally repeats words, has difficulty comprehending spoken language, writes with a lot of inaccuracies and misspellings, and gives the wrong names for words, they may suffer from a problem originating in Wernicke's area. Other results may lead to the conclusion the damage is in the occipital region, Broca's area, or some other language-related center of the brain. Unfortunately, in most cases the damage cannot be repaired; however, speech therapy often leads to significant improvement and, in many instances, a complete return to normal speech.

The Least You Need to Know

◆ No one knows the origin of speech, but you can take your pick between the biblical explanation and a series of scientific theories.

◆ Your brain learns to recognize words, grammatical principles, and concepts, and young children, especially, can pick up foreign languages.

◆ For most people the left hemisphere of the brain is the language center, but if it is damaged, the right hemisphere can sometimes take over this function.

◆ Approximately one in five Americans have some form of reading disability, and about 5 percent of Americans suffer from speech disorders.

Chapter 10

Making Sense of It All

In This Chapter

- The camera in your head
- Listen closely
- Aromas of all kinds
- A matter of taste
- Feeling touched

When you were a kid, you learned that humans have five senses: vision, hearing, taste, smell, and touch. Later, you may have heard references to a "sixth sense," a power of perception that does not require use of the other five. Some scientists now argue that humans have many more senses, perhaps as many as 20. These senses include very specific abilities to detect gravity, electricity, and ultraviolet light. Nevertheless, this chapter focuses just on the basic five.

Mind's Eye

How important are your eyes? Three quarters of all the information your brain will process today will come from your eyes. Yet your eyes don't really see anything; your brain is what provides the gift of sight. For example, if your visual cortex is damaged, you won't be able to see even if your eyes are perfect.

The eyes are balls of tissue that contain transparent jelly, which keeps them round. Each eyeball is typically just under an inch long. The white of the eye, the *sclera*, is a coating that covers the eyeball. Six small muscles allow the eyeball to look straight ahead and swivel to look up, down, and to the side.

The eye is covered by the clear *cornea* and gets its color from the amount of a pigment called *melanin* in the *iris* (meaning "rainbow"), which produces only four colors: blue, green, gray, and brown. The dark hole in the center through which light enters the eye, is called the *pupil*.

The human eye is typically compared to a camera because a number of its components perform camera-like functions. The pupil acts like the camera aperture by controlling the amount of light allowed into the eye; more light is allowed in when light is dim and less is allowed in when light is bright. The lens of both the eye and camera focus the light. When a person looks at faraway objects, the muscles around the lens relax so the lens becomes thinner. The muscles contract, making the lens thicker, when the person looks at close objects. The back of the eye is composed of a layer of cells that are sensitive to light and change, like film, when they are exposed. This layer of cells is the *retina*.

In Living Color

The retina has two special types of cells: cones and rods. Cones work best in bright light and help a person see the clarity and color of the stimulus. The seven million cones in the eye are capable of receiving only three colors: red, blue, and green. We can see other colors when a combination of these cone cells is triggered. Believe it or not, the human eye can detect 10 million different shades.

Light enters the cornea, then passes through the pupil and lens to the retina.

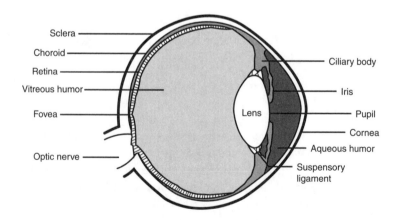

Rods are far more numerous than cones. The 125 million rods in your eyes are responsible for detecting brightness. Together rods and cones convert light into a pattern of nerve signals that are transmitted to the brain.

Code Blue

When you are driving, the blind spot is the place in your field of vision where you can't see what is behind or beside you. Each eye also has a blind spot, a round area on the retina where the optic nerve enters the eye and where there are no rods or cones. This area is not sensitive to light and is therefore "blind." These blind spots don't normally hinder vision because each eye compensates for the blind spot in the other eye.

Chiasmic

The image formed on the retina is upside down because of the way the lens bends the incoming light rays. Visual information travels along the optic nerve to the neuronal track switching station called the optic chiasm. At this point, fibers from the nasal (or inside) half of each retina cross to the opposite side of the brain; fibers from the temporal (toward the ear) half of the retina do not cross. The principal receiving station for the information carried by these fibers is the visual center of the cortex in the occipital lobe.

Code Blue

When a tumor of the pituitary gland grows large enough to compress the optic chiasm, it causes an inability to see in the periphery of the visual field. For example, a person with this condition could drive down the middle of a one-way street without difficulty, but the person would not see the cars parked on either side of the street.

From there, more specific signals are transferred to other regions of the brain. For many years, scientists believed that objects transmitted visual codes that were printed onto the retina and then the visual cortex decoded the message as a whole image. Late in the twentieth century, however, new research showed that the neuronal impulses from the eyes are broken up into components, so the message relating to color goes to one area, the signals for horizontal lines go somewhere else, and the transmissions related to where the head is in space are processed in a third area. Meanwhile, the signals associated with perception and recognition go to the higher centers of the cortex to be analyzed and compared with stored memories. The brain also corrects the orientation of the image so that it is turned right side up.

A Deer in the Head's Lights

To get a sense of how the vision process works in real life, imagine that you are hiking through the woods and see an animal rustling in the bushes. The cranial nerves cause your eyes to constrict and focus on the area where the disturbance is coming from; they also carry messages from the brain telling your muscles to move your head to follow the movement of the bushes. The cerebrum calculates the size and speed of the animal and compares the images that are coming through the eye to memories of similar shapes, colors, and movements. The possibility that the animal is a bear stimulates the basal ganglia, cerebellum, and motor cortex to raise the body's level of awareness. The heart is told to pump faster and the lungs are directed to take in more oxygen. When the animal comes out of the bushes and the brain recognizes the visual image as a harmless deer, the muscles are given the message to relax.

Two Eyes Are Better Than One

Humans have two eyes that move together. (In contrast, a chameleon's eyes can move independently, enabling the lizard to see in two different directions simultaneously.) This *binocular vision*, as vision that uses two eyes is called, has distinct advantages over a single eye.

Two eyes can see a larger area than just one, a total field of vision of about 200 degrees. The overlap between what the right and left eye can see is a bit narrower, 140 degrees, and this overlap defines the field of our 3-D vision. This field of vision helps us to judge distance accurately.

It's All a Blur

Everyone is familiar with the Snellen chart with the big "E" on top and knows that a passing grade on their eye test is 20/20. But do you know what this grade means? This grade refers to the lowest row of letters on that chart that a person with normal vision can read from a distance of 20 feet. If you score worse than 20/20 vision, say 20/40, you don't really fail, but this score means that when you are 20 feet away, you can read a row of letters that someone with normal sight can read from 40 feet away. Oddball Americans use this 20/20 standard. Most of the world uses the metric system where the goal is to have 6/6 vision, a number which refers to meters rather than feet.

In addition to the Snellen test, you probably have taken the Ishihara test for color blindness in which you are asked to pick out an image from a circle of colored dots. The individual with normal vision can pick out a specific number from the dots, but the color-blind person sees a completely different number in the circle. Someone who is color blind cannot see the normal range of colors and may see some as gray. The most common form of color blindness is red-green, which means an inability to distinguish between reds, greens, and browns. Interestingly, color blindness is far more common in men than women, but no one knows why.

Color blindness can't be cured, but glasses can correct vision problems indicated by the Snellen or other eye tests. Unclear images result when the lens of one or both of the eyes does not properly focus light rays on the retina. For example, distant objects look fuzzy to a near-sighted person because their lenses direct light to the front of the retina. Wearing glasses with concave lenses focuses the light rays on the proper area of the retina, and the image is made clearer. The need for corrective lenses may be obviated altogether by the use of newer surgical procedures such as *LASIK*.

Even perfect vision is likely to deteriorate with age. The elasticity of the lens of the eye tends to diminish over time, which makes it more difficult to focus. The cornea can also change shape and become less transparent, causing blurred or distorted vision. More than half of Americans 65 and over suffer from *cataracts*. Cataracts are cloudy areas of the lens, which decrease the amount of light that passes through the lens and bends entering light abnormally. The effect is to cloud vision, which may make it more difficult to read a book, drive a car, or see clearly. Today, cataract surgery is one of the most successful surgical procedures.

Words of Wisdom

LASIK is the acronym for laser-assisted in situ keratomileusis. In LASIK surgery, precise and controlled removal of corneal tissue by a special laser reshapes the cornea and thus changes its focusing power. Before deciding on this procedure, a person should consult with one or more eye surgeons.

Seeing Is Believing?

Sometimes the brain makes mistakes. Yes, it's true. Sometimes we see something, and it is not what our brain tells us it is. Look up at the moon on any clear night. How big does it look? Depending on its position in the sky, it may look enormous and so close you can reach out and touch it. Other times it is a small, distant sphere. The size of the moon, of course, hasn't changed. When the brain interprets the way you see an image to make it fit a known pattern, one that does not match the actual image, the result is called an optical illusion.

You've probably seen lots of these illusions. For example, you can draw two parallel lines of equal length. If you draw arrows at the end of one line facing in and arrows on the other line that face out, the lines no longer look as though they are the same length. There are also lots of pictures that you can look at and see different images. In one of the most popular, you may see either an old woman or a young girl—or both. Maybe you've also tried one of those tricks where you stare at an object or a rotating spiral and then look at a blank wall and you "see" the object there. The Dutch graphic artist M.C. Escher created some of the most artistically interesting illusions such as impossible buildings, repeating geometric patterns, and other pictures that challenge our ability to interpret visual images.

An Ear for It

The ear is divided into the outer ear, the middle ear, and the inner ear. The outer part is the most visible, but least important; it's little more than a flap of skin and cartilage (the auricle) on either side of the head whose primary purpose is often to attract ridicule. The outer ear does collect sound waves and direct them into an approximately one-inch canal (the auditory canal) that leads into the head before hitting a dead end at the eardrum (the tympanic membrane for the cognoscenti), a thin piece of skin the size of the fingernail on your pinky. Changes in air pressure cause the eardrum to vibrate.

> **Gray Matter**
>
> One reason that many animals can hear better than humans is that their outer ears are more efficient, moving like radar antennae to pinpoint the direction of sound. The animal with the keenest hearing is the bat. Vampire bats can hear frequencies more than eight times greater than a human, which makes it easy for them to find humans and drink their blood (just kidding!).

Beyond the eardrum is the middle ear, an air-filled cavity that contains the body's three smallest bones, the ossicles, which are better known as the hammer (malleus), anvil (incus), and stirrup (stapes). The hammer is about three-tenths of an inch long and could easily fit on the tip of your finger.

The visible part of the ear has little to do with hearing. The main components that allow us to hear are deeper inside the head.

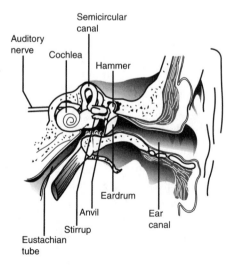

A tube (the eustachian) extends from the middle ear into the back of the throat. The eustachian tube allows air in and out of the ear so the pressure inside the ear is equal to that outside of it. If the pressure is unequal, as it sometimes is when you are descending in an airplane, the eardrum is unable to vibrate properly.

Vibrations flow past the eardrum and pass through the ossicles, which each add to the vibrations' intensity so that the force of the vibrations is ultimately increased by 20 to 30 times by the time the vibrations reach the inner ear.

The inner ear contains three semicircular canals: the cochlea, utricle, and saccule. The cochlea is the organ that is directly involved in hearing. This 1.5-inch, snail-shaped tube spirals nearly three times. Within these twists is a thin membrane covered with more than 20,000 sensory hair cells, each of which has 50 to 100 hairs. Sound vibrations cause ripples in the fluid in the tube, which moves the membranes and the tiny hairs, causing the hair cells to transform the sound waves into nerve signals, which are then transmitted to the auditory cortex in the brain via the cochlear nerve and other nerves. This part of the brain alone has 100 million neurons.

Each hair cell can fire up to 20,000 nerve signals per second, meaning that 400 million sound related messages should reach the brain every second. It doesn't happen that way, however, because only 30,000 nerve fibers are available to carry the signals, and they each can handle only 1,000 signals per second or a total of 30 million messages. This difference means a great deal of information is lost.

As with the visual information from the eyes, most, but not all, of the fibers carrying the sound from the ears cross over to the opposite side of the brain.

The incoming messages are compared by the two hemispheres and referenced against the catalog of sounds in the memory. Based on this comparison, the brain gives the body instructions on whether to take any action and decides whether to retain the information in the memory. How this process happens remains a mystery.

The hearing process is quite remarkable. Think about how you can filter out certain sounds. For example, a mother may be able to sleep through anything else, but if her child stirs in another room, she's awake immediately. Her brain is constantly monitoring sounds and somehow can distinguish what is important.

One other interesting phenomenon about hearing, as well as the other senses, is that different people perceive the same stimulus differently. For example, the young, bright, and incredibly handsome coauthor of this book has the good taste to enjoy rock music while his old curmudgeonly father considers it to be horrific noise.

Good Vibrations

The vibration of molecules creates sound. Sound is measured in *decibels*. The softest whisper is approximately 10 decibels, which is the quietest sound humans can hear. When sound reaches 130 decibels it becomes painful, but anything above 90 decibels can damage hearing. Experts recommend that you protect your ears with earplugs or some other covering when noise exceeds 85 decibels, which is just above the din of heavy traffic. Government regulations prohibit workers from being exposed to noise above 115 decibels and restrict how much time a worker can be subject to lower noise levels. The following table lists the decibel levels of common sounds.

Common Sound Levels	
20 dB	Whisper
35 dB	Quiet rural nighttime
40 dB	Quiet home
70 dB	Ordinary conversation
90 dB	Heavy traffic
100 dB	Subway train
120 dB	Jet airplane/discomfort
130 dB	Pain threshold
150 dB	Air-raid siren

The human ear can detect only a limited range of frequencies, which are measured in *hertz*. Humans can hear sounds ranging from 15 hertz to about 20,000 hertz while a dog can pick up sounds at 30,000 hertz, which is why a dog whistle is beyond our hearing but not our pets' (bats can hear up to 100,000 hertz). Keep the human range of hearing in mind the next time that you're shopping for stereo equipment and the salesperson tries to talk you into buying equipment with a frequency range that is beyond your ability to detect.

Words of Wisdom

Sound intensity is measured by **decibels** (dB). Sound frequency is measured in **hertz** (Hz), which is the number of cycles of vibrations per second.

The ear can detect the differences between 40,000 different sounds. Interestingly, detecting sounds from the left or right is easier than detecting sounds from up or down because of the shape of the ears, which are better suited to "catch" sound waves from the side. Sound waves travel at 750 miles per hour, and the brain can immediately determine which ear receives a sound first and thereby judges the direction of the sound.

Sound Off

The ear is a very sensitive organ that people typically abuse by listening to loud music, working in noisy environments, and tolerating noise pollution. Ears are resilient, but repeated exposure to loud sounds can cause permanent damage. Cranking up your stereo to hear Led Zeppelin's *Whole Lotta Love* at what you once considered the appropriate volume may have been satisfying and a good way to drive your parents crazy, but it also destroyed thousands of your high-frequency neurons. Those poor folks who stand on the tarmac at the airport directing airplanes to their gate and who are constantly bombarded by jet noise probably face a similar loss of hearing, despite wearing protective earmuffs.

An estimated 10 million Americans have some hearing loss, and millions more are at risk. In addition to deafness, many people with damaged ears are afflicted with *tinnitus*, or ringing in the ears.

Profound hearing loss is usually divided into two types: perceptive and conductive. People suffering from perceptive hearing loss have some physical defect in the inner ear or a problem with the connection between the auditory nerve and the brain. If the condition is congenital or caused by damage later in life, the loss of hearing is likely to be permanent. The amount of loss can range from partial to total deafness and may be partially corrected by the use of a hearing aid or more dramatic measures, such as cochlear implants.

IQ Points

Cochlear implants are a relatively new form of treatment for profoundly deaf individuals whose auditory nerves remain functional. Electrodes are surgically embedded in the cochlea to stimulate the auditory nerve and are connected to a receiver surgically placed beneath the skin. A microphone near the ear relays sound signals to a microprocessor, which converts them into electric signals that are sent to a transmitter behind the ear and on to the receiver and cochlear electrodes. Although the cochlear implant does not reproduce the human voice, the device substantially improves sound perception for many users, particularly those children and adults who became deaf after they had learned to speak. They are less effective, but still helpful, for many people who became deaf before learning to talk. In 2001, popular radio personality Rush Limbaugh lost his hearing from an autoimmune disease and had cochlear implant surgery, which his doctor called a success.

When sound vibrations can't reach the inner ear, people suffer conductive hearing loss. In some cases, the cause is an object blocking the ear canal or a buildup of wax. Removing the blockage cures the problem. Infections can also cause hearing problems when inflammation in the ear prevents the normal vibration of the eardrum and ossicles. These conditions sometimes go away by themselves or can be treated with antibiotics.

A possible cause of unilateral (one-sided) hearing loss is the presence of a tumor of the acoustic nerve at the base of the brain (in the cerebellopontine angle if you want to know the precise location). This tumor may be treated by microsurgery with a team made up of a neurosurgeon and a neurootologist. In recent years, these tumors have been treated with the use of the gamma knife (a noninvasive source of pinpoint radiation therapy). Finally, the aging process also leads to a gradual degradation of the ability to hear.

On the Scent

From the sweet aroma of a flower to the stench of garbage, the brain of someone with a particularly sensitive schnozz, such as a chef or wine connoisseur, can interpret as many as 10,000 different smells while the rest of us have to make do with about 3,000. People's noses don't vary much in sensitivity; the difference is in a person's ability to concentrate on smells and identify and remember those that are familiar. The sense of smell is also an adaptation needed for survival. The ability to smell rotten food, for example, helps protect us from eating something dangerous.

So how exactly does a person identify a smell? Much of the process is a mystery, but some of the pieces have been uncovered.

The nostrils lead into a cavity located above and behind the roof of your mouth. Each nasal cavity has a patch of tissue that is smaller than your thumbnail and contains about five million olfactory cells. Like the cells in the cochlea, these cells have tiny hairs, six to eight per cell, which detect the chemicals that cause smells.

As in the case of hearing, dogs smell much better than humans. Humans have 40 million olfactory receptor cells, but a dog has a whopping 200 million cells. A dog's nose is also better equipped to pick up smells because it is (usually) large and wet. You may have thought the wetness was just a trait of adorability, but it's actually one of *odor*ability because it collects and dissolves the scent particles more efficiently than our dry noses. Come to think of it, a dry nose and a poorer sense of smell isn't such a bad tradeoff.

The information about the incoming odors is instantly translated into a nerve signal. The nerve signal then travels about an inch to reach the area of the brain that interprets it.

Though we don't know for sure, the suspicion is that the sense of smell works a little like the ability to distinguish color. Remember that the cells in the retina can receive the light waves of just three colors, but the combination of those wavelengths make it possible to see a rainbow of colors. Scientists have identified seven basic scents (some believe the number may be as high as 30), and the range of possible smells may be a combination of them. These smells are decay, flowery, musk, peppermint, ether, spicy, and mothballs. (With that list, it's a wonder that people want to take a whiff of anything.)

This idea means that you probably don't have a specific receptor in your nose for the smell of a rose or a skunk. Still, the brain is somehow able to distinguish between the two. Neurons from the nose pass through relay stations called the olfactory bulbs and onto the thalamus, which in turn is connected to the frontal cortex where signals are compared to those in memory for recognition. Other nose neurons are linked to the limbic system, which is concerned with motivation, emotion, and certain types of memory. The fact that the interpretation of smell is in part performed in this part of the brain helps explain the association people often make between smells and specific memories and feelings.

People can detect thousands of smells, so how do they keep their senses from being overloaded with scents? The answer is that the brain filters odors just as it does sound. Just as we can get accustomed to a sound and not be aware of it, so too can we become desensitized to a particular smell. Then again, there's no getting used to some odors.

> **Code Blue**
>
> One of the authors of this book and his mother suffer from anosmia, the absence of a sense of smell. This condition can occur for reasons that are unknown, but it is usually a result of an infection, tumor, or the overzealous use of nose drops.

Yum and Yuck

The sense of taste is closely related to the sense of smell. As with vision and smell, there are basic tastes: bitter, sweet, salty, and sour. These basic tastes can combine to form more than 10,000 flavors. Special cells on the upper surface of the tongue called taste buds detect these tastes.

The cells for each specific taste are not spread evenly over the tongue; they are in particular locations. The tip of the tongue is the site of the taste buds for sweetness. Further back on the sides of the tongue are the receptors for saltiness. Even further back is the location of the sour taste buds, and the back of the tongue is the site of bitterness. Not all taste buds are on your tongue. Some are in the throat, inside the cheeks, and on the roof of the mouth as well.

> **IQ Points**
>
> Scientists increasingly accept the existence of more than four basic tastes. The taste called umami is associated with the chemical glutamate, which is the "G" in MSG that is often used in Chinese food. The taste is rich and meaty. Researchers also suggest that fat may be a taste as well. Of course, most people have already figured out that fat stimulates the taste buds.

Taste buds allow us to differ-entiate among foods. The four types of taste buds— salt, bitter, sour, and sweet— are distributed across the tongue in specific areas, so the entire tongue does not "taste" the same thing.

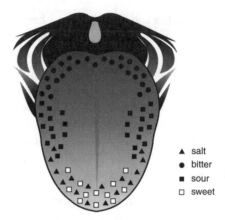

▲ salt
● bitter
■ sour
□ sweet

The description of the distribution of the taste buds is the one that appears in just about every book on the senses, but new research suggests that this picture is probably not accurate. The individual cells of a taste bud may actually respond to all tastes, with the distinctions being made when the signal from the receptors in the tongue reaches the brain.

As in the case of cells involved in the other senses, specific molecules are apparently associated with particular receptors, much like a key fitting into a lock. In this instance, the molecules from food hold the key to particular taste buds, which, in turn, produce an electrical signal that is transmitted to the limbic system and cerebral cortex.

The motor centers of the brain instruct the muscles to lift a fork to the mouth, chew, and swallow and direct the glands to salivate. The limbic system is the pleasure center that gives you the feeling that you either liked or hated what you've tasted. The frontal cortex than sends a message to the speech center that prompts you to say, "Delicious!" or in the case of our kids, "Ick!"

Gray Matter

Why do most people love chocolate? Some scientists think there's more to it than just its great taste. They believe that chocolate may trigger the release of certain chemicals in the brain that produce a feeling of pleasure. Others believe chemicals in chocolate may mimic the effects of marijuana by affecting chemicals in the brain that cause feelings of happiness. Nothing definite has been established, but this is definitely the kind of research you can sink your teeth into.

Like vision and hearing, the sense of taste declines with age. People start out with about 10,000 taste buds, but by the age of 60, they're down to fewer than 7,000, which is why older people often have a more limited sense of taste and tend to like spicy foods.

How Touching

The sense of touch is actually a group of senses: the ability to detect cold, heat, pain, and pressure. If you compare the size of the areas of the brain devoted to the individual senses, you'll find that the touch center is significantly larger than the small regions devoted to smell and taste.

The skin is the largest sensory organ, covering about 20 square feet in an average adult (and weighing in at 6 pounds!). The receptors, however, are not on the outside of the skin, as you might expect. A thin protective layer of skin that can't feel anything covers the body. The receptors for touch are in a second layer of skin. The areas with the most dense concentration of receptors, such as the fingertips and lips, are the most sensitive. Sensitivity to pain also varies for different parts of the body. For example, it requires very little pressure on the eye to cause pain, but a great deal more to hurt the palm of your hand. When any of these receptors is stimulated, it sends a message to the brain.

Touch is another of the body's defense mechanisms. The feeling of pain can warn of damage or danger. If we feel something that is very hot, we learn to pull away. If we sense the skin being cut, we stop what we are doing. Some diseases interfere with this sense. People who suffer the now rare condition known as leprosy do not feel pain or heat and therefore may not recognize when they are in danger of being burned.

Drugs, such as morphine, can also minimize the sense of pain. In addition, the brain can sometimes block out pain by itself by the release of the body chemicals called endorphins. For example, under conditions of extreme stress, such as during an athletic competition or combat, athletes or soldiers can often ignore the immediate pain they feel to win a race or save a comrade.

> ## IQ Points
> Body hair plays a role in controlling temperature. When receptors in the skin detect a drop in body temperature, the hairs stand straight up. This reaction traps air between the hair and the skin, helping to insulate us against the cold and causing the sensation referred to as goose bumps.

The behavior of athletes and soldiers also illustrates the fact that individuals perceive pain differently. For example, one person may take an aspirin or some over-the-counter medicine to control pain whereas another may require a prescription drug for the control of the same pain. This is particularly true of people with chronic problems such as back pain.

People also have different responses to pain depending on the situation. Say you are hanging a picture on your wall and hit your thumb with a hammer. This accident causes severe pain and swelling in your thumb, which may last for 48 to 72 hours. Twenty-four hours later, you have a date to go to lunch with your mother-in-law, and you cancel because the pain in your thumb is so severe. That same morning your best friend calls and says he has an extra ticket to the NFL playoff game. You forget about the pain in your thumb and go to the football game.

Incidentally, the brain itself cannot feel anything because it has no touch receptors. Neurosurgeons can therefore operate on the brain with the patient fully awake, which is particularly important in the removal of brain lesions in critical areas such as those controlling speech and arm and leg movements.

The Sixth Sense: More Than a Movie

Besides being a great suspense film, the sixth sense is thought to be a power of perception that is separate from the other senses. It could be nothing more than intuition, or, as some believe, it could be a form of extrasensory perception.

Usually included in this category of phenomena are telepathy, or thought transference between persons; clairvoyance, or supernormal awareness of objects or events not necessarily known to others; and precognition, or knowledge of the future. Plenty of people have claimed these powers, and they are often given legitimacy by the press.

A number of skeptics have offered rewards to anyone whose claims of paranormal ability can be demonstrated before a panel of experts. A number of people have taken up the challenge, but, to date, no one has proven they have a sixth sense.

The Least You Need to Know

- Your eyes can't see; they only transmit visual information to the brain, which must then interpret it based on the memories of colors, shapes, position, and other features of the image. Sometimes the brain is fooled, resulting in an optical illusion.
- The human ear is very sensitive and can identify and filter thousands of sounds, but it is also easily damaged, and hearing tends to deteriorate with age.
- The sense of smell is interpreted in part by an area of the brain associated with emotion, which may explain the connection people make between feelings and certain odors.
- The taste buds, in combination with the sense of smell, enable you to recognize thousands of flavors from a combination of just four, or perhaps five or six, basic tastes.
- The touch center of the brain is relatively large and is crucial to the body's ability to protect itself from harm.
- Though it makes for great science and horror fiction, no one has produced any credible evidence for the existence of ESP.

Chapter 11

Life's Necessities

In This Chapter

- ◆ Feeling hungry?
- ◆ Sleep is good
- ◆ Waving goodnight
- ◆ Can't help dreaming
- ◆ Sex is very good

We know that the history and anatomy have been scintillating, but it's time to focus on what really interests most of us: food, sleep, and sex. As explained in this chapter, our most primitive instincts are biological needs that are controlled by the brain.

Eat, Drink, and Be Merry

Researchers have found that the brain has centers that inhibit and stimulate eating. Oddly enough, they're both in the same general place, the hypothalamus. To be precise, the *ventromedial* area of the hypothalamus curbs the appetite while the *lateral* hypothalamus encourages us to pig out.

How do we know this?

Glad you asked. With apologies to the animal rights folks, this knowledge comes from animal experiments. Researchers have found that electrically stimulating the ventromedial area of the hypothalamus inhibits eating and that destroying this part of the brain causes voracious eating and, ultimately, obesity in rats. Similarly, damage to the lateral hypothalamus makes an animal lose its appetite and starve to death while stimulating the area can cause the animal to overeat.

The desire to eat and continue eating is also influenced by the sense of taste. We like to eat when we are served something that tastes good. Surprise! On the other hand, the pleasure we derive from taste usually dissipates the longer we eat, which helps discourage us from overdoing it.

Research shows that obese people tend to eat fewer, but larger meals, but no one knows exactly why they do this. The research suggests that obesity may be related to abnormalities in the hypothalamus.

> **Words of Wisdom**
>
> **Anterior** means toward the front. **Lateral** is the direction away from the midline. **Ventromedial** refers to the lower part in the center or midline.

Normally, the balance of water in the body determines thirst. If we don't have enough water, we either drink, or the brain signals the kidneys to decrease water loss in urine by increasing water reabsorption into the blood. Electrical stimulation of the *anterior* hypothalamus induces drinking and damage to it, stops drinking. Removal of the lateral hypothalamus can cause abstinence from eating and drinking.

Sweet Slumber

Without sleep we fall apart. We get ornery and nod off in the middle of activities, and if deprived of sleep long enough, we can hallucinate.

If personal experience isn't persuasive enough, consider the Darwinian perspective. Caveman and woman were vulnerable to attack from predators while they were sleeping, which meant sleep was a potentially harmful habit for the human species. Yet our ancestors apparently enjoyed a reasonably good night's sleep, so rest must be important. When you consider that a third of your life (25 years for a 75-year-old person) is spent sleeping, it had better be worth the lack of effort!

> **Code Blue**
>
> A lack of sleep, especially REM or deep sleep, will kill a person more quickly than lack of food. In 1965, Randy Gardner set the world record for time without sleep by staying up for 264 hours (11 days).

As in the case of eating and drinking, a particular part of the brain, the midbrain reticular formation in this case, has been associated with wakefulness. Studies have shown that stimulation in this region can wake up an animal and damage to this region can put the animal into a coma.

We know that a part of the brain determines whether we are awake, but do we sleep because the wake center somehow gets tired or is there a sleep area of the brain that somehow turns off the wake center? Research suggests the existence of a sleep center, but it is spread across several parts of the brain, including the thalamus and the pontine reticular formation.

You've Got Circadian Rhythm

Some people seem to get by on a few hours of sleep a night, yet others seem to need regular naps or 12 hours of sleep a night to fully function. As your mother probably told you when you were a kid, most of us need about eight hours of sleep. Typically we need less sleep as we age. Infants require as much as 20 hours a day, a six-year-old needs about 10, and a 65-year-old requires just 6.

IQ Points

Somnambulism, better known as sleepwalking, is a mostly unexplained behavior. Apparently, the parts of the brain that control movement and speech remain awake and send signals to get up, walk, talk, and perform other actions that give the impression that the sleepwalker is no longer asleep though he or she is. Interestingly, the sleepwalker can usually navigate around the house without bumping into things, but may not recognize the difference between a garbage can and a toilet. Waking someone who is sleepwalking is difficult, if not impossible, and is generally not a good idea. The sudden shock of being awakened can cause fear and disorientation. When the person does wake up, he or she does not remember having done anything unusual during the night. Sleepwalking is more common among children than adults. In fact, 30 percent of children between the ages of 5 and 12 have walked in their sleep at least once, but even those who do it regularly usually outgrow the habit.

The alternating pattern of sleeping and waking occurs in a regular pattern over a normal 24-hour period. This *circadian rhythm* is an internal body clock based in the pineal gland, which is keyed to daylight and darkness and tells us when to hit the sack and when to rise and shine. The system is more complex than a clock because the brain keeps track of cumulative rest so that the body automatically catches up on sleep if a person doesn't get the amount needed during a particular cycle. This internal timer is found in two clumps of nerve cells in the hypothalamus, an area called the suprachiasmatic nuclei.

Your Brain Does the Wave

An electroencephalograph (EEG) records the waves of electricity generated by brain cells. Electrodes are attached to the scalp to detect the waves, and the EEG produces a line on a paper similar to what you see in movies with lie detectors to indicate the changes. These brain waves vary with changes in the body; for example, they are slow if we are relaxed and conscious and fast if we're stressed. Brain waves come in four types:

- Alpha waves oscillate about 8 to 13 times per second and are seen during wakefulness and a period of relaxed consciousness; they disappear during deep sleep.
- An intense state of alertness, as in stressful situations, produces beta waves, which fluctuate from 13 to 30 cycles per second.
- Delta waves appear during deep sleep when the wave cycles slow to less than 4 per second.
- Drowsiness or a light sleep results in theta waves, which oscillate 4 to 7 cycles per second.

Cycling Through Sleep

If you watch people sleep, you might see them squirm or hear them mumble from time to time, but mostly they just lay there unconscious. But there's more to sleep than meets the wide-awake eye!

Sleep may give the body a chance to rest, but the brain is no less active during sleep than it is during wakefulness. During sleep, the brain gives the body instructions for growth, healing, and learning.

Sleep consists of a cycle of four stages that progress from light to deep sleep. The cycle lasts for about 60 to 90 minutes and at the end of each new cycle, the period of dreaming lasts longer. The following table describes each of the stages in the sleep cycle.

The Stages of Sleep	
Stage 1	Heartbeat and breathing slow; muscles relax. It's easy to be awakened.
Stage 2	EEG shows a pattern of "sleep spindles," eyes roll slowly, and loud noises are required to wake the sleeper.
Stage 3	This stage is characterized by long, slow delta waves. Heartbeat, breathing, blood pressure, and body temperature all drop. Muscles relax.
Stage 4	After 20 to 30 minutes of sleep, EEG shows mainly delta waves. During this time of deep sleep, talking and sleepwalking may occur.

In Stage 1, breathing and heart rate slow down, and brain waves are low voltage and irregular. The pineal gland releases the hormone melatonin, which, in turn, leads to the production of the neurotransmitter serotonin, which is related to sleep.

After a few minutes, Stage 2 begins, and your eyes roll from side to side and effectively turn off. Even if your eyes were open, you wouldn't be able to see. Brain waves become more irregular.

As your body continues to relax, you enter Stage 3, and body functions slow down further. Brain waves become more even, large, and slow.

Finally, in Stage 4, deep sleep is reached, and the EEG indicates long, slow delta patterns. Then the cycle repeats, but in reverse, going from Stage 3 to Stage 1. Only this time, Stage 1 includes dreaming.

Perchance to Dream

From ancient times, humans have had a fascination with dreams. For much of history, many people believed that dreams were messages from gods or the afterlife. Rulers would make policy decisions based on their dreams or their advisers' interpretations of them. In the Bible, Joseph's ability to correctly interpret the Egyptian pharaoh's dreams helped him escape prison and become a prince (Genesis 41).

Daydreams occur when people allow their attention to drift and fantasize or imagine things, but the dreams that take place during sleep are a different story.

Dreaming occurs at a particular point in the sleep pattern, when the EEG activity (the pattern of brain waves) in the cortex is fast with a low amplitude. During this stage of sleep, the eyeballs move as if they are following an object; hence, the stage is called REM (rapid eye movement) sleep the rest of the stages are called NREM or nonrapid eye movement sleep. Although the eyes are active, the rest of the body is paralyzed, a contrast from the other stages of sleep when people may toss and turn. One theory is that the muscles are inactive during REM sleep to prevent people from acting out their dreams.

Just as we require sleep, we also appear to have a need to dream. If our sleep is interrupted so we don't have the proper periods of REM sleep, we will make up for the lost dreams later. The first period of REM sleep is short, about 10 minutes, but the periods of REM sleep become progressively longer as the night goes on, and the final one may last as much as an hour. Typically, a person spends about two hours each night dreaming.

IQ Points

While Freud had begun to study dreams a half century before in his therapy sessions, the physiological study of dreams did not really begin until the 1950's after a graduate student at the University of Chicago discovered REM sleep after attaching electrodes to his son.

Gray Matter

Freud argued that traumas and repressed emotions are buried in the unconscious but come out in dreams. He believed in trying to remember fragments of dreams to help uncover the buried feelings. Freud focused on the idea that repressed sexual feelings lead us to dream of fulfilling our desires. In his view, the anxiety surrounding these desires turns some dreams into nightmares.

But why do we dream? That question remains unanswered, though some researchers and psychiatrists believe that there are links between dreams and some forms of mental illness and that dreams can reveal information about the subconscious mind. Neurologists see them as more mundane physiological processes. The pons secretes the chemical acetylcholine, which stimulates the cortex, and dreaming ensues. Another part of the brain stem produces noradrenaline to activate REM sleep. Some scientists believe dreams are nothing more than the random firing of neurons.

No one knows what really occurs during dreaming. Perhaps the brain uses this time to file and sort memories, recording some and discarding others.

Wishing You Could Sleep

Do you

- Have difficulty falling asleep?
- Wake up too early in the morning?
- Wake up frequently during the night and have difficulty going back to sleep?
- Feel tired even after sleeping?

Code Blue

Narcolepsy is a disabling illness marked by a permanent and overwhelming feeling of sleepiness and fatigue. Other symptoms involve abnormalities of dreaming sleep, such as having dreamlike hallucinations and feeling physically weak or paralyzed for a few seconds. The disease affects more than 1 in 2,000 Americans, and most are never diagnosed or treated.

If your answer to any of these questions is yes, you may have insomnia. Transient or short-term insomnia may occur one night or over the course of a few weeks. If the problem occurs periodically, the insomnia is considered intermittent. When sleeplessness recurs most or every night for a month or more, the problem is chronic.

Insomnia can be caused by everything from the temperature of the bedroom to the food eaten before lying down to stress. It tends to more often affect people over age 60, women, and those with a history of depression. Chronic insomnia can also be caused by physical or mental disorders, such as asthma, Parkinson's disease, and arthritis. It may also be related to the misuse of alcohol, caffeine, or other substances.

Insomnia can be treated. Treating an underlying physical or psychological problem may ameliorate the sleeping difficulties. Sleeping pills are sometimes prescribed, but they can have side effects and are supposed to be used only for a relatively short time. Relaxation techniques, such as self-hypnosis, can be used to reduce stress that leads to insomnia. Some people spend too much time trying to fall asleep and can improve their chance of success by doing other things and restricting the amount of time they do sleep, so they will be more tired in later nights. Yet another approach, which is often used with young children who don't have insomnia, but simply have trouble sleeping, is to restrict the use of the bed to sleep (and sex for adults). This restriction is designed to condition the mind to associate bed with sleep and no other activity.

The novelist Charles Dickens described obstructive sleep apnea in 1836. You may not have heard of it, but sleep apnea is as common as adult diabetes, today affecting more than 12 million Americans. Obstructive sleep apnea is caused by a blockage of the airway, usually when the soft tissue in the rear of the throat collapses and closes during sleep. With each apnea event, the brain briefly arouses the sufferers from sleep in order for them to resume breathing; consequently, sleep is extremely fitful and of poor quality. Untreated, sleep apnea can cause high blood pressure and other cardiovascular disease, memory problems, weight gain, impotency, and headaches. Sleep apnea can be diagnosed and treated.

> **CAUTION**
>
> **Code Blue**
>
> One frightening symptom of narcolepsy is sleep paralysis, an abnormal occurrence during REM sleep where a patient suddenly finds himself unable to move for a few minutes, most often upon falling asleep or waking up.

Sleepless in Seattle and Other Cities

Air travelers are familiar with another type of sleep disorder, jet lag. This disorder occurs when you travel across time zones, and your body clock is thrown out of whack when it experiences daylight and darkness at the "wrong" times. Symptoms include fatigue, early awakening or insomnia, headache, constipation, irritability, and reduced immunity. The symptoms are generally worse after you've flown in an easterly direction and often persist for a day or more while you adjust to the new time zone.

You don't even need to get on an airplane to suffer from jet lag. The same problem can occur by staying up all night or working night shifts or rotating shifts over 24 hours (such as the schedule of medical interns).

No one has found a sure-fire way to avoid jet lag, but suggestions abound. Scientific research, for example, has suggested that shining a light on the back of your knees during flight stimulates a chemical process that resets your internal clock. An over-the-counter vitamin B supplement is also being tested as a jet lag pill. In addition, many people feel that taking the hormone melatonin in the evening in the new time zone can significantly decrease or eliminate jet lag. British Airways offers a variety of tips for minimizing the effects of jet lag, such as drink plenty of water during the flight, consume coffee when you need a lift rather than all day, and don't exercise at bedtime because it might wake you up rather than tire you out.

Jet lag is not just exhausting and frustrating, it may also have more serious long-term effects. Recent research suggests that chronic jet lag may be dangerous because it can interfere with memory and damage the temporal lobe. This research is a cause of concern for frequent travelers, pilots, and flight attendants.

Sex Is a Drive

We made you wait until the end of the chapter for it, but now we're ready to discuss sex. Though you may doubt it, the truth is that unlike eating, drinking, and sleeping, sex is not necessary for your survival. It is, of course, a prerequisite for the survival of humanity, but any one of us could make it through life without any sexual activity.

The hypothalamus stimulates the gonads to release sex hormones. The gonads, in turn, send their own hormones to the hypothalamus, which initiates sexual behavior. Researchers in animal labs have found that they can stimulate parts of a rat's hypothalamus and drive it into a sexual frenzy during which it will forgo food and starve in favor of the chance to excite itself.

Gray Matter _____

A recent study suggested that music can be as stimulating as food or sex. The study, conducted at McGill University, found that musicians who reported getting the "chills" from pieces of music had activity in the parts of the brain, in particular the midbrain and areas of the cortex, that are associated with the feelings of euphoria common from food and sex. Like so many of these dramatic-sounding news reports, this one was based on a very small, select sample of subjects, and the relationship between music and pleasure is hardly proven by the results; nevertheless, it sounds good to us.

The way we experience sex is entirely a product of the mind. The sensations are transmitted from receptors in the erogenous zones to the brain. In addition, the brain is the source of our memories about what is pleasurable, the place where fantasies are produced, and where our feelings toward our partners are generated. The brain can also be the source of problems related to sex, such as anxiety related to performance, fear of pregnancy or disease, and shyness or feelings of shame.

One of the most controversial aspects of brain research relates to sexual orientation. Between 1 and 5 percent of men and women are homosexual in every culture. Is this because there are always some people who prefer this lifestyle or is it because a certain percentage are wired that way? Some people believe that orientation is determined by choice or is learned; others insist that it is biologically determined. Because everything psychological can be said to be simultaneously biological, sorting out the causes may be impossible.

This type of research also produces chicken-and-egg arguments. For example, one researcher found that a cluster of cells taken from the hypothalamus of dead people who were known to be gay and straight were different for the two groups. The cluster was larger in heterosexual men and smaller in women and homosexual men. This difference suggested a biological influence on orientation, but some psychologists argue that sexual behavior may alter the brain's anatomy just as experiences during childhood change the brain's wiring.

For now, the debate continues.

The Least You Need to Know

- The brain has centers that inhibit and stimulate eating.
- Everyone needs to sleep, and the body has an internal clock that makes sure we get as much as we need.
- A variety of factors can cause sleeplessness. A common one is jet lag, which new research shows can be overcome. More serious are disorders that cause insomnia. These disorders can sometimes be cured by eliminating external factors such as stress or noise; other times they require medication or other therapeutic measures.
- Dreams are an important part of normal body function. Dreaming takes place at different intervals throughout the night during REM sleep.
- The brain stimulates the release of hormones that influence the sex drive. The basis for sexual orientation may also be linked to biological differences in the brain, but this theory remains a matter of controversy.

Chapter 12

The Body's Autopilot

In This Chapter

- The body's thermostat
- Fight or flight
- Reflexively speaking
- Potty training
- Sex mechanics

When you switch on the thermostat in your house, you know that the temperature in the house will be automatically regulated, and you probably don't worry about how this task is accomplished. You just know that if your house becomes too hot or too cold, your thermostat will adjust the amount of heat or cooling necessary to keep your house at your chosen temperature.

Our bodies also have a thermostat, which is located in the hypothalamus. (As you may remember from Chapter 5 the hypothalamus is under the cerebrum and thalamus and in between the hemispheres of the brain.) The hypothalamus regulates body temperature without offering us any conscious choice about the setting; it also regulates a variety of internal processes without us having to think about them by controlling what is called the *autonomic nervous system*. This system handles the automatic actions of the body that function at the subconscious level. In contrast, the other half of the nervous system, the somatic nervous system, is the voluntary part (mainly muscles) that we consciously control.

How important is the autonomic nervous system?

Think about all the different things that are constantly on your mind: work, children, money, groceries. Now consider how much more difficult life would be if you also had to worry about how fast your heart should beat, whether digestion should start or stop, or when to make your pupils dilate. And how would you accomplish these bodily functions while you were sleeping? Those day-to-day external worries don't seem quite so overwhelming now, do they? This chapter examines the autonomic nervous system and describes how the body takes care of itself.

Your Internal Thermostat

When all our internal settings are normal, we say the body is in equilibrium or *homeostasis*. This state is constantly changing, however, as we experience alterations in environmental temperature or stress, eat food, or shift our posture. When any change occurs, the autonomic system, which is part of the *peripheral nervous system* located outside the brain and spinal cord must restore or adjust the settings. The system regulates the body's settings by using the nerves that stimulate these three major types of tissue: smooth muscle, cardiac muscle, and glands. *Smooth muscle* refers to those muscles in the stomach, intestines, bladder, eye, skin, and around blood vessels. The cardiac muscle is the heart.

The autonomic system can be further broken down into the sympathetic and the parasympathetic nervous systems, which, as their names suggest, often work in opposition to each other.

Yet another subdivision of the autonomic system is the enteric nervous system, which does not directly relate to the brain. It is a collection of neurons in the gastrointestinal tract that are related to digestion. It also includes neurons in the pancreas and gall bladder.

Words of Wisdom

Smooth muscles appear smooth when looked at under a microscope and are associated with internal organs. In contrast, skeletal muscles, which, strangely enough, are controlled by the skeletal nervous system, are attached to bones and looked striped under a microscope.

In Chapter 7 we described in detail the motor neurons and how they are "connected" by synapses. In the skeletal system, the neurons follow an uninterrupted path between the muscle and the central nervous system. In the autonomic system, however, an intermediary is involved. One group of neurons synapse outside the central nervous system with another set of neurons, which then connect to the muscles. The synapse where these two groups of nerves come together is called a *ganglion*.

The neurons coming from the central nervous system to the ganglion are referred to as preganglionic neuron. The neuron that gives rise to the preganglionic neuron is situated in the gray matter of the spinal cord. Those neurons that travel from the ganglion to the muscles are called postganglionic neuron. Communication across the synapse between the preganglionic and postganglionic neurons is carried by the neurotransmitters acetylcholine and norepinephrine.

The sympathetic and the parasympathetic nervous systems share this basic ganglion structure; however, the systems differ in the details. For example, the preganglionic neurons in the parasympathetic system originate in the brain stem and the lower part of the spinal cord (the sacral region). In the sympathetic system, however, the preganglionic neurons sprout from the thoracic and lumbar regions of the spinal cord, that is, the thoracolumbar region, and extend from the base of your skull to the coccyx (tailbone).

The size of the neurons also differs in the two systems. In the sympathetic system, the preganglionic neurons are short, and the postganglionic neurons are long. In the parasympathetic system, the opposite is true.

Words of Wisdom

The autonomic system is part of the **peripheral nervous system**, meaning it is anatomically outside the brain and spinal cord, although physiologically it is intimately related to the brain and spinal cord. This system has 36 pairs of peripheral nerves; 31 pairs are spinal nerves that enter the central nervous system below the neck, and 5 pairs are cranial nerves that connect directly to the brain.

Finally, the preganglionic neurons of both systems release the neurotransmitter acetylcholine. Neurons that release acetylcholine are called *cholinergic*. The difference between the systems is in the chemicals that are released by their postganglionic neurons. Whereas postganglionic neurons of the parasympathetic system are cholinergic, the postganglionic neurons of the sympathetic system are *adrenergic*, which means they release norepinephrine and epinephrine (better known as adrenaline). The neurons for the sweat glands, which are cholinergic, are the only exceptions.

These differences between the sympathetic and parasympathetic systems are important because they are the reason that the two systems of nerves have different effects on the same organ. For example, the parasympathetic stimulation of the stomach increases peristalsis (the muscle contractions that move food through the digestive system) and relaxes sphincters whereas sympathetic fibers have the opposite effect.

Energize!

You probably know about the autonomic nervous system from what's commonly called the "fight-or-flight" response. This response occurs when the body senses an emergency, such as when a ferocious-looking dog begins barking and moves menacingly in your direction. While your cerebrum considers whether you should stand still and take your chances with the beast or make a run for it, your sympathetic nervous system rockets into action, expending energy to make your heart beat faster, increase your blood pressure, and slow the digestive process to allow you to take action you voluntarily decide upon. Under strong emotional circumstances, sympathetic and parasympathetic fibers also may discharge, causing involuntary emptying of the bowel and bladder, which means that you are so scared that you wet and soil your pants.

The autonomic nervous system is divided into the sympathetic and parasympathetic systems, which act as on/off switches for a variety of body functions.

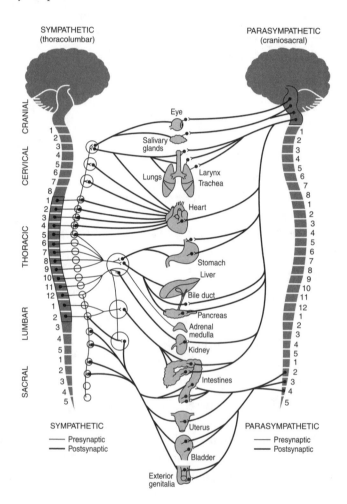

The body also takes steps to automatically respond to less urgent situations, such as the need to digest food. Another such response is the contraction of the blood vessels to maintain pressure when we get up in the morning so the blood doesn't concentrate in our lower body and cause us to faint. An abnormality in this part of the autonomic nervous system may produce a condition known as postural hypotension, which is characterized by a rapidly falling blood pressure when a person gets up too quickly. This condition may cause fainting.

A mere change in external temperature sets off a chain reaction of automatic events. A rise in body temperature, for example, is registered by the hypothalamus, which transmits a message to the sweat glands to do their thing and cool down the body. The blood vessels dilate to increase the flow of blood to the surface of the body, from which body heat radiates. You may not want to sweat, but you can't do much about it except exercise your volitional control and take a shower to cool yourself off.

Stressed Out

When we think of a fight-or-flight response, probably the first situation that comes to mind is a serious physical threat, but other forms of *stress* can trigger this response as well. Mental stressors can include an inability to solve a complex or an interpersonal problem at work, general anxiety, or depression. Stressors can also be physiological, such as pain, fatigue, and chronic infections. A less obvious cause of stress may be environmental factors, such as pollution, crowds, and noise.

Words of Wisdom

European physician Hans Seyle defined **stress** as "the non-specific response of the body to any demand made upon it." He called the "demands" stressors.

When the body encounters stress, the sympathetic system is activated. The postganglionic neurons of this system release adrenaline. The adrenaline then stimulates nerves, which in turn stimulate tissues throughout the body. As a result, the gastrointestinal tract reduces its activity, heart rate and blood pressure increase, coronary arteries dilate, and the breathing tubes called bronchials in the lungs dilate. Damage to the sympathetic system in the neck causes an interesting condition known as Horner's syndrome, which is characterized by processes called meiosis, ptosis, and anhydrosis. In this syndrome, the pupil is constricted (meiosis) because the muscle that allows the pupil to dilate is paralyzed. Drooping of the eyelid (ptosis) occurs because of the denervation (a disconnection of the neural pathway) of the smooth muscle that raises the eyelid. However, the eyelid can still be raised voluntarily through the action of the skeletal muscle fibers in the eyelid mediated by cranial nerve III. Absence of sweating (anhydrosis) makes the side of the face and neck with the damaged sympathetic system appear reddened and feel warmer and drier than the other side.

Overloading the System

If the body is placed under too much stress, it can break, which is why so many health problems are associated with stress. Hans Seyle explained the process in terms of three steps: alarm, resistance, and exhaustion.

The immediate physiological response to stress is alarm, which is the process described previously by which the body comes to a state of maximum alertness. Ordinarily, stress dissipates after this immediate alarm state, and the body automatically returns to its normal state.

If the stress is more acute or prolonged, however, the body has to work harder and enters a resistance stage where the pituitary gland is stimulated to release adrenocorticotropic hormone (ACTH), which, in turn, signals the adrenal gland to release a class of stress hormones known as corticosteroids. Over time, some of these hormones can damage the body. For example, they may affect the production of insulin, which can cause diabetes, or affect the viscosity (thickness and stickiness) of blood, which can result in high blood pressure and increase the risk of strokes.

The key to escaping the resistance stage is to reduce or eliminate the cause of stress. The body can still return to normal, but the longer this stage continues, the more damaging it is to your health.

If the stress does go on too long (and how long too long is differs for everyone), the body reaches the point of exhaustion. It can't fight back any longer, and the person may suffer serious depression and even degenerative diseases.

CAUTION

Code Blue

Why do people get ulcers from too much stress? Stress causes an increase in histamine production in the body, which causes an increase in gastric acid secretion and ulcer formation. Ulcers also can develop from the treatment of severe head injury, especially those cases complicated by coma, and in all cases of increased intracranial pressure such as brain tumors) where steroids are used to combat the pressure. Steroids increase gastric acid secretion, which ultimately may cause ulcers.

Cool Down

Certain causes of stress are beyond our control, such as the polluted environment, a death in the family, or a car accident. Still, we can take steps to reduce the stress in our lives and even to minimize the effect of those uncontrollable factors. Here are a few suggestions:

- Use relaxation techniques, such as deep breathing, meditation, positive visualization, and yoga.
- Take a 10 minute walk and also exercise regularly.
- Get lots of sleep.
- Volunteer and help others.
- Think about what you need to get done and try to work smarter not harder.
- Amuse yourself with jokes and television or movie comedies.

- ◆ Allow yourself to relax with free time.

- ◆ Find a physical outlet for stress, such as punching a punching bag, squeezing a ball, or taking a karate class.

- ◆ Avoid cigarettes, coffee, and alcohol.

- ◆ Share the cause of your stress with others so you don't have to face it alone and keep it bottled up inside.

The Conservationist

The parasympathetic system primarily controls glands and other visceral organs that work together to conserve and restore the body's energy. When you are relaxed, for example, the parasympathetic system causes the heart to beat slower, blood pressure to decrease, and the digestive process to begin.

In individuals with peripheral vascular disease (partial or complete obstruction of one or more blood vessels in the legs), doctors may perform an operation called a lumbar sympathectomy (removal of the sympathetic ganglia) to reduce the sympathetic influence on the blood vessels. This operation allows parasympathetic action to dilate the blood vessels, thereby permitting more blood to flow to the extremity.

Now that you're an expert on the cranial nerves (see Chapter 8, if you need a refresher course), you'll understand when we say that cranial nerves III, VII, IX, and X play important roles in the parasympathetic system. One example of a parasympathetic function is the oculomotor nerve's (III) regulation of the iris and lens of the eye. If a bright light shines in your eye, the sensory neurons send a message to the midbrain, which answers with a command via the fibers in cranial nerve III to contract the pupillary muscles so the pupil will become smaller and reduce the amount of light allowed to enter the eye.

Another example of a function of the parasympathetic system is the regulation of the secretory glands that supply saliva to the mouth, mucus to the nose, and tears to the cornea of the eye. (You've heard of blood, sweat, and tears; well, this is spit, snot, and tears!)

The following table shows how the sympathetic and parasympathetic systems operate in different parts of the body. As you can see, the effects of each are almost polar opposites.

IQ Points
The principal neurotransmitter involved in the sympathetic and parasympathetic systems is acetylcholine. The actions of this chemical in the autonomic system are similar to those of nicotine and therefore called nicotinic.

The Autonomic Nervous System

Body Part	Sympathetic Stimulus	Parasympathetic Stimulus
Stomach	Peristalsis reduced	Gastric juice secreted, motility increased
Heart	Heart rate and force increased	Heart rate and force decreased
Eye muscle	Pupil dilation	Pupil constriction
Lung	Bronchial muscle relaxed	Bronchial muscle contracted
Liver	Increased conversion of glycogen to glucose	
Kidney	Decreased urine secretion	Increased urine secretion
Bladder	Wall relaxed, sphincter closed	Wall contracted, sphincter relaxed, bladder emptied
Small intestine	Motility reduced	Digestion increased
Large intestine	Motility reduced	Secretions and motility increased
Adrenal medulla	Norepinephrine and epinephrine secreted	
Salivary glands	Saliva production reduced	Saliva production increased
Mouth and nose	Mucus production reduced	Mucus production increased

Knee-Jerk Reactions

You accidentally put your hand on the burner of a stove that's been turned on. Pain shoots up through your hand, and you yank it away from the heat. Interestingly, this reaction is not caused by your brain saying, "Hey, your flesh is burning, so you should move your hand off the stove." In fact, your brain is not involved at all in the instantaneous reaction to the feeling of heat on your hand. This reaction is known as a *reflex*. It is an immediate, involuntary response mediated by just two neurons, one afferent (sensory) and one efferent (motor), connected in the spinal cord.

How do we know what we did to stimulate the reflex? That's where the brain comes in. The reflex involves an impulse that rebounds from the receptors in your hand along the nerves to your spinal cord and back. Meanwhile, the spinal cord passes the message up to the brain, which interprets what happens, identifies the type of pain (burning) and the object that caused it (a hot burner), and reflects on the ill-considered decision that brought about the pain.

A slightly different situation occurs when you step on something sharp. If your right foot, for example, steps on a glass shard, it instantly flexes upward, and the opposite leg extends due to excitation of one group of motor neurons and inhibition of another group simultaneously. Put simply, your body reflexively knows to make adjustments to maintain balance. You don't even need to step on glass to see how this reflex works. If you lift up one foot, the other will tense; otherwise, you'd fall over.

Reflexes are important for a number of reasons, including providing a means for the body to quickly avoid danger when the brain is otherwise engaged.

Consider what would happen if the brain had to react to the pain of your hand on the hot stove. What if you were concentrating on a telephone conversation or watching TV and it took a few extra seconds to recognize you were on fire?

Think this sounds silly? You're sure that it couldn't happen? Well, how many times have you burned something you were cooking because your brain was occupied with some other activity?

Baby Steps

We are born with most of our reflexes, and some of them are unique to infancy. For example, if you touch a baby's cheek, he or she will open his or her mouth and turn to try to find the nipple of a breast or bottle. Another reflex, called the Moro reflex, involves the throwing out of a baby's arms and legs when the baby falls backward or is startled. And have you ever noticed that when you hold babies up to try to get them to walk, they make stepping motions with their legs even when they're still newborns?

In adults, the protective impulse often involves multiple reflexes. For example, if someone throws something at your head, your eyes will shut, the skin of your face will tense, your neck muscles will twist your head to the side, and your arms will shoot up in front of your face.

Gray Matter

The knee-jerk test is one way a doctor can check whether your nerves and muscles are working properly. To perform the test, the doctor taps just below your kneecap with the rubber hammer. That action stretches the tendon, which spits out a nerve impulse that travels along the sensory nerves to the spinal cord, which responds along the motor nerves with a message to the muscles in the front of your thigh to contract, causing the lower leg to jerk forward.

Another example you may recall from childhood is the game you played with your friends to try to make the other person blink. You might have pretended to poke their faces or clapped your hands in front of their eyes. The reflex is to blink in response to these actions. When you see someone make a threatening movement toward your eye, impulses from the retina travel to the midbrain, the reticular formation, and the nuclei of the facial nerves, which then send a message to activate the muscles of the eyelids and cause them to close. Clapping produces a response in the auditory nerve, and then, through a connection in the midbrain, an impulse travels to the postganglionic fibers to the eye, causing you to blink. These responses are instantaneous.

Holding It

Infants reflexively empty their bladders and bowels, but as they mature they learn to control this reflex. And boy are parents glad when they do!

How do kids learn to control their bladders? After a baby is born, the myelination of nerve fibers (the development of the covering of the nerve fiber that promotes conduction) extending to the bladder continues over a significant period of time (24 to 36 months). The ability to control the bladder depends on the completion of this myelination, as well as the completion of fiber tracts that extend from the cortex of the brain to the sacral area of the spinal cord and the time it takes for babies to gain experience to understand what parents want. One-year-olds can't control their bladders because their nervous systems haven't matured enough. But during the maturation process, children are taught the proper potty procedure by their parents, and eventually the combination of the maturation of the nervous system and the association of the feeling of discomfort with the need to go to the bathroom leads children to the milestone that gets them out of diapers.

You might want to hold your nose for this part, but we do want to briefly explain how you relieve yourself. First, the bladder fills with urine, which stimulates receptors in the bladder that suggest to the midbrain the need for a stimulus to cause the bladder to contract and the sphincter to open. Parasympathetic nerves control urination. *Proprioreceptors* in the wall of the bladder respond as it fills and stretches. When enough urine accumulates in the bladder (approximately 200 to 300 cubic centimeters), you become aware of the need to go to the bathroom.

> **Words of Wisdom**
>
> **Proprioreceptors** detect stimuli from the muscles and tendons and help us orient our limbs and bodies.

Really Letting Go

Although many people may insist otherwise, sex is not a reflex. To initiate a sexual response, conscious effort is necessary to send messages from the cerebrum through the thalamus to the autonomic nervous system. The usual four stage process of sexual response involves multiple levels of brain activity. Both the parasympathetic and the sympathetic divisions of the autonomic nervous system are required for sexual activity.

The first stage is usually arousal. This may be stimulated by different senses, such as the touching of erogenous zones, pleasing smells, and the sight of a partner's body. The brain raises the body's temperature and rushes blood to the sex organs. As activity continues, heart and respiratory rates accelerate. This is the second stage. The autonomic system kicks in now as the body moves on to the third stage.

In males, the parasympathetic neurons are responsible for producing an erection. Sympathetic neurons initiate the contractions needed to ejaculate, but ultimate ejaculation is caused by parasympathetic activity. In the case of females, vaginal secretions, erection of the clitoris, and engorgement of the labia minora are caused by parasympathetic neurons.

Ejaculation and genital contractions are associated with intense feelings of pleasure, which we call an orgasm. This powerful emotion derives from a transfer of impulses from the sex organs through the thalamus to the cortex where the sensation of pleasure is realized. The last stage is one of relaxation after the orgasm as the body returns to its normal level of activity.

Kinda takes the thrill out of sex, doesn't it?

Diseases affecting the parasympathetic and sympathetic systems can cause sexual dysfunction. For example, diabetes affects parasympathetic neurons and can cause impotence and the inability to ejaculate. Diseases affecting the sympathetic neurons can also impair ejaculation.

The entire sexual response is mediated by the hypothalamus and the limbic system. These parts of the brain influence many aspects of our emotional lives, such as anger, rage, placidity, fear, and social attraction (we'll have more to say about this topic in Chapter 15). The role of the limbic system helps explain why emotions can affect sexual function. For example, they can cause over stimulation of the sympathetic system, which can result in weakness of erection and premature ejaculation. Certain types of drugs, such as antidepressants, sedatives, antihypertensives, and antipsychotics can cause impotency or a decrease in or total loss of sexual interest or arousal. Damage to the hypothalamus or the limbic system or multiple areas of the cortex may have the same effect. Brain damage caused by head injury, strokes, toxins, illicit drugs, or alcohol abuse may also have similar effects.

The Least You Need to Know

- The body has its own internal thermostat to regulate vital functions.
- The sympathetic and parasympathetic systems automatically turn body processes on and off.
- Reflexes involve neuronal circuits that don't require the involvement of the brain.
- We can learn to control some autonomic activities, such as the need to relieve ourselves.

Chapter 13

It's Your Choice

In This Chapter

- ◆ Staying upright
- ◆ Sensing the labyrinth
- ◆ Moving too fast
- ◆ Clocking reaction time
- ◆ Quick thinking

Most of the last chapter dealt with strictly involuntary body functions and actions. A number of other activities, however, involve a combination of the involuntary and voluntary nervous systems. As this chapter explains, sometimes we have to think about what we're doing while parts of our brains make other decisions on their own.

On Balance

Clap your hands.

Now close your eyes and clap your hands again.

You made a voluntary decision to clap, but without seeing what you were doing, you still knew that your hands would come together. To give a more complex example of this kind of mental process, consider a gymnast. She may make the conscious decision to walk on a balance beam, but she will not decide whether her foot lands in the right place, allowing her to stay on the beam. The difference lies in the fact that walking on the balance beam involves two different parts of the brain. Her cerebrum makes the voluntary decision, and her cerebellum mediates the activity. (Remember the baseball example from Chapter 6?)

The brain puts together disparate pieces of a behavioral puzzle to create knowledge. In the case of the gymnast, the receptors in the muscles, called proprioceptors, detect the expansion and contraction of the muscles and send the information to the brain, which uses it to determine the position of each part of the body.

When you think about it, you may recognize when you are off balance, but ordinarily you do not consciously think, "I need to shift my weight to keep from falling down." No, your brain automatically calculates the movement required to keep you upright. However, different activities require different levels of conscious thought. For example, you don't need to think about balance when you're walking, but if you are learning to ski, you do need to make a conscious effort through the cerebrum to shift your balance to keep from falling down.

More Than Five Senses

Remember in Chapter 10 when we said you had more than five senses? One of these extra senses is the labyrinthine sense, which helps you keep your balance. It is aided by yet another sense, the kinesthetic sense, which allows you to be of the position of your limbs; it works with the help of the receptors in the joints and ligaments.

One way doctors can test for damage of the labyrinthine receptors is to check whether patients can keep their balance with their feet together and eyes closed. If a patient sways and falls to one side, this indicates an injury and is called Romberg's sign (after a nineteenth-century neurologist named Moritz Romberg).

The No-Look Pass

Blind people can learn to be great pianists. How can they do this without the visual cues needed to find the keys? Kinesthetics, my friend.

The joints, the juncture between your bones, contain receptors that become deformed when your limbs move. These deformations translate into nerve impulses, which travel to the brain and relay information about the angle of your arms and legs. The rate at which the neurons fire also relates the speed at which the joints are moving.

The kinesthetic information is carried from the joints to the spinal cord and via two tracts (the funiculus cuneatus, and funiculus gracilis if you're interested) to the medulla. From there the message is sent to the midline structures of the cerebellum, then to the thalamus, and ultimately to the primary somatosensory area of the cortex.

So how do blind people master the piano? It takes a combination of practice and experience. The pianists hear the notes and transpose what they hear to the notes on the keyboard. They probably don't get the notes right on the first try. The pianist might play a recording of "Moonlight Sonata," sit at the piano, and then whack a few keys. The sound may not come out right at first, but if the person has learned the association between particular keys and notes, he or she can perform and perhaps master the sonata after some trial and error. The better a person knows the keys and notes, the more readily that person can play a composition.

> **IQ Points**
>
> If you play piano, you can see what we are talking about by playing a recording of a piece of music that you are unfamiliar with and then trying to play it on your keyboard with your eyes closed. If you are an accomplished musician with a good ear, you'll quickly learn the piece without having to ever peek at the keyboard.

Your A-Mazing Sense

Stand. Sit. Walk. Run. You probably have never given much thought to how you do these simple things. Each of these actions requires your labyrinthine sense to maintain your balance.

Consider someone engaged in a more complex activity, such as gymnastics. A gymnast's body is in almost constant motion: somersaulting, jumping, falling, and landing. The brain must calculate the position of the body at each moment, measuring the acceleration and rotation, commanding the muscles in the legs to relax when tumbling and tense on landings, and directing the arms to extend to stay straight and the eyes to fix on a takeoff and landing spot. It's pretty darn amazing that the brain can do all this instantaneously without you giving it a second thought.

Perhaps you're wondering why the sense we've been talking about is called labyrinthine. The answer is that the detection of the relationship between head movements and body movements is performed in the vestibular system, which is headquartered in a maze-like series of bony tunnels and canals in the inner ear known as the bony labyrinth.

> **Gray Matter**
>
> The utricle responds to gravitational forces and linear acceleration chiefly in the horizontal plane (that is, moving forward/backward under the affect of gravitation). One way to understand linear acceleration is to think of the gravitational forces of a jet aircraft catapulted off the deck of an aircraft carrier.

The bony labyrinth basically has two functions: the detection of the head in motion in space and the detection of the position of the head (and body) at rest in space. How does the bony labyrinth detect motion?

Inside the bony labyrinth is the membranous labyrinth. Perilymph (a fluid) fills the space between the two. Endolymph (another fluid) fills the space inside the membranous labyrinth. Vestibular hair cells are found within the membranous labyrinth as well. The membranous labyrinth has two swellings (or sacs), the utricle and saccule, and three semicircular canals: the anterior, the horizontal, and the posterior. Body movements cause the fluid to move in these sacs, and this fluid movement generates a neural impulse.

These nerve impulses from the inner ear travel along the vestibular part of cranial nerve VIII to the medulla. The medulla may trigger a response in the stomach and cause nausea (see the following section). Some impulses go on to the oculomotor nerve to regulate eye movements to compensate for head movement. Others go to areas of the brain related to motor coordination.

Turning Green

Motion sickness is caused by a disconnect between what the inner ears tell the brain about what the body is doing and what the eyes transmit that it is doing relative to the ground and space. For example, if you are on a roller coaster when it loops upside down, your eyes and ears both send the message that the body has flipped over. If, however, you are doing something as passive as reading a book on a ship that is bouncing up and down with the waves, the eyes remain fixed on the page, but the ear registers the movement of the rest of the body with the churning seas. The combined signals cause you to feel dizzy, tired, and nauseated. The signal may then go out to the vagus nerve to evacuate your lunch. (This could never happen with our book!) Motion sickness is a manifestation of prolonged and excessive stimulation of the vestibular system.

Some people are especially susceptible to motion sickness from an early age. Children suffer more from motion sickness than adults, and by age 50 most adults no longer have this problem. But even those adults who have motion sickness are not alone if they're on an airplane that experiences heavy turbulence. In those conditions, one third to one half of the passengers are likely to suffer some degree of motion sickness. The likelihood of getting sick on a plane is increased by the fact that fear and anxiety can lower the threshold for experiencing symptoms.

> ## IQ Points
>
> Astronauts are sometimes thought of as having the toughest constitutions. After all, they are able to go on those centrifugal contraptions that spin around in a circle at high speed and tolerate the acceleration and turbulence of a rocket launch. Still, they often get space sick because weightless conditions in space send conflicting cues to the brain from the ears and eyes.

An Ounce of Prevention

Unfortunately, there are no sure-fire ways of preventing motion sickness. Some over-the-counter medications, such as Dramamine, can help and sometimes keep the symptoms in check if the motion is not too severe. Prescription drugs that can be worn as patches behind the ears or armbands on the wrists are available as well. Like other drugs, these drugs can have side effects, so consult a doctor and read warnings before taking any medication. Some natural herbs and other remedies are reputed to prevent motion sickness or minimize its effects.

In addition to pharmaceutical approaches, you can try to put yourself in positions and areas of moving vehicles that minimize the sense of motion. For example, the front seat of a car, the interior of a ship, or the front of an airplane is more stable than other locations. Always choose forward-facing seats if you are given a choice. Try to watch the horizon when you're on a boat and get plenty of fresh air. Don't try to read! You're much better off sleeping, if you can. Also, many people think it is better not to eat before traveling because it will keep you from vomiting, but a light meal is actually a better idea because it generally helps settle the stomach. Snacking on salty pretzels or crackers during the trip is another good idea.

IQ Points

Choose a large ship for your next cruise. The larger the ship is, the less likely it is to give rise to conditions that cause motion sickness. Most modern cruise ships are equipped with stabilizers to decrease motion in rough seas. Similarly, when flying, try to fly on a big plane. In bad weather, a 747 moves a lot less than a Piper Cub, which bounces like a rowboat in the ocean.

The Room's Spinning

Of course, you can also just become dizzy without getting motion sickness. Remember when you were a kid and just spun around in circles until you got dizzy and started to stumble around the room? What happened was that the fluids in your inner ear moved as you spun, and when you stopped, the fluid continued to spin.

A variety of factors can cause dizziness. The most common causes of dizziness are problems with blood circulation. If your brain doesn't get enough blood flow, you feel light-headed. (A complete loss of blood supply to the brain causes unconsciousness in 8 to 10 seconds and death within seconds after that.) In other instances poor circulation is a result of heart disease or problems with arteries such as high blood pressure, arteriosclerosis (hardening of the arteries), or blockages. Nicotine from cigarettes and caffeine in coffee or tea can decrease the blood flow to the brain, as can some other drugs. Too much salt can also affect circulation.

Gray Matter _____

A different kind of "motion sickness" is vertigo, which is the sensation that the world is spinning around you or that you are revolving in space. Vertigo is usually marked by dizziness caused by a problem, such as an infection, in the inner ear and may also cause feelings of giddiness, faintness, and light-headedness. However, just being dizzy does not mean that you have vertigo.

A cold or viral infection can affect the inner ear and may cause dizziness. If the infection is severe enough, it may damage the inner ear and do more permanent damage to hearing and equilibrium. More serious neurological diseases and injuries to the inner ear can also cause serious or permanent hearing or equilibrium impairment.

A not uncommon condition known as Meniere's disease is a serious condition affecting the vestibular system. It is characterized by a sudden attack of severe vertigo, nausea, vomiting, tinnitus (ringing in the ear), and eventual unilateral deafness (deafness in the affected ear). A surgery that cuts the vestibular portion of cranial nerve VIII has been a successful form of treatment for difficult cases of this condition that don't respond to less intrusive forms of therapy.

The Pyramids

The brain is the master of the three types of muscles in the body: the skeletal, smooth, and cardiac. The skeletal muscles, which allow us to move our arms, legs, fingers, and other bones, are controlled by neurons that descend from the cortex to the spinal cord by the extrapyramidal and *pyramidal systems*.

The cerebellum governs balance through the extrapyramidal tracts. These neurons monitor the tension and position of the muscles and relay this information to the brain where the cerebellum determines whether it is necessary to alter the motor commands or change the position and tone of the muscles.

Words of Wisdom _____

The **pyramidal system** gets its name from the pyramid-shaped neural bundles in the area of the medulla that most of the pyramidal tracts cross.

The cortex also plays a critical role in ensuring that the muscles relax. Experiments on animals have found that cutting off the brain stem from the spinal cord renders them paralyzed. Muscle reflexes continue to function, but the animal has no voluntary control. If a part of the brain stem is left intact, the animal's limbs become stiff and are unable to bend. This result suggests that the higher brain centers are involved in muscle tone and prevent us from becoming rigid.

The extrapyramidal system also prevents the muscles from becoming overstimulated. If your body didn't have this control, the doctor testing your reflexes with a little tap on the knee might get a foot in the face, and you might end up with serious tendon damage.

The frequency of impulses in the motor neuron determines the degree of muscle contraction. To prevent overloading the system, the body allows the motor neurons to conduct impulses only at a certain frequency. Like most everything else we've discussed, this self-protective process is complex, but one aspect worth mentioning is the sensory receptors known as Golgi tendon organs (yes, the same Golgi who discovered the nerve cells). Whenever a muscle contracts, it stretches the Golgi tendon organ. If the stretch is great enough to exceed the threshold that poses a danger to the body, the Golgi tendon organ triggers an impulse that inhibits the transmission to the nerves that are causing the muscle to contract.

If the extrapyramidal system malfunctions, the result can be serious. Parkinson's disease, for example, is linked to damage to this system, which leads to postural rigidity and uncontrolled tremors.

Penfield's Map

The neurons in the extrapyramidal tracts synapse in multiple places in the brain (collectively called the basal ganglia), but the pyramidal neurons travel directly to the motor neurons in the spinal cord that control skeletal muscles and synapse there.

Through brain mapping experiments, scientists such as Penfield and others have found that the cortical areas devoted to different muscles are systematically arranged. For example, the motor area of the cortex that controls the thumb is next to the area that controls the index finger. As we explained in Chapter 4, the size of the area of the cortex responsible for a particular function is related to the sensitivity of that part of the body. (Remember the homunculus?)

Gray Matter

Electrocorticography (intra-operative recording from the surface of the brain) can determine the locus or origin of a seizure disorder, which allows a surgeon to pinpoint the area that needs to be removed.

We also know that one hemisphere of the cortex usually controls the motor function on the opposite of the body. Motor function occurs when the precentral gyrus is stimulated. Stimulating the supplementary motor area that sends impulses to several muscles can cause more than one action. Such stimulation might cause a patient to lift her arm and turn her head to look at it.

The fact that the brain itself has no pain receptors allows neurosurgeons to operate within the brain with the patient awake. Neurosurgeons can therefore determine the effect of stimulation or probing on a specific area of the cortex, such as a vocal response (from the speech area) or a motor response (the movement of one of the extremities, for example). Patient awareness during surgery is extremely important in surgery for epilepsy and for the removal of brain tumors.

Put the Hammer Down

When the nervous system detects a stimulus and the body responds automatically, the response is called a reflex. When we have a chance to make a decision as to what to do, however, the response is called a reaction. The basic difference between reflex and reaction is the involvement of the brain; the brain is not involved in reflexes.

Reactions cover a wide range of activities, but all basically involve the consideration of what to do. In some cases, you may have a long time to react. In a game of chess, for example, you can study the move of your opponent for some time before deciding on the appropriate move, which involves a mental evaluation of the options as well as a physical movement of the piece on the board. At other times, you must react in a split second.

Consider a hockey goalie's plight. Professional hockey players can hit a puck in the neighborhood of 90 miles per hour. The goalie must recognize the shot has been taken, determine the puck's speed and direction, and move his or her arm, leg, or both to block it all in less than half a second.

IQ Points

A hockey puck is the fastest-moving object in any major professional team sport. The hardest shot, the slapshot, has been clocked at more than 100 miles per hour. If this number isn't scary enough, consider that the hardest servers in men's professional tennis can hit the ball more than 120 miles per hour and the fastest serve is reputed to be 138 miles per hour by American Steve Denton. Even this number pales in comparison to the speed of the fastest game of all, jai alai, in which the record for the fastest speed is 188 miles per hour.

Most of us don't have the formidable task of stopping a Brett Hull slapshot, but we do have a far more dangerous challenge on the highways. If you're traveling at 55 miles per hour behind a pickup truck just 165 feet in front of you and something falls out of the truck, you have less than two seconds to figure out what's on the road, where it is, and what you should do to avoid it. Scary as that may sound, it's still better than the Old West when the person who was second best at drawing their gun ended up dead.

Now Draw!

Think you're fast?

Before donning a goalie's mask and standing in the net, check your reaction time with a homemade test. Get a yardstick and have someone hold it from the top so that it is perpendicular to the ground and several feet above the floor.

Hold your thumb and index finger about an inch apart opposite the 18-inch mark. Tell your friend to let go of the stick without giving you any warning and then try to catch it before it falls. Then check the inch mark where your fingers grab the stick. Subtract this number from 18 or subtract 18 from the number to see how many inches the stick fell before you caught it. You can determine your reaction time with the following table.

Distance Dropped	Reaction Time
2 inches	0.10 seconds
4 inches	0.14 seconds
6 inches	0.18 seconds
8 inches	0.20 seconds
10 inches	0.23 seconds
12 inches	0.25 seconds
14 inches	0.27 seconds
16 inches	0.29 seconds
18 inches	0.31 seconds

The average person's reaction time is 6 to 8 inches, about .18 to .20 seconds. A goalie trying to stop a 100 mile per hour shot from 20 feet away would need to have a reaction time of about 4.5 inches or .15 seconds.

From Physical to Mental

It's been a long and somewhat complex story, but we hope you now have a pretty good understanding of the relationship between the brain and physical activities. We've covered the senses, voluntary and involuntary behaviors, language, and our basic needs. Now it is time to go deeper inside the brain to explore what are sometimes referred to as the higher processes related to thinking, learning, memory, and emotion. As David Letterman would say, this is more fun than human beings are supposed to have!

The Least You Need to Know

◆ The labyrinthine sense helps us keep our balance without having to think about it.

◆ Receptors in our joints give us a kinesthetic sense that allows us to know the position of our limbs.

◆ Motion sickness occurs when the inner ear tells the brain that the body is moving one way and the eyes signal it is moving a different way.

◆ Conscious decision-making distinguishes a reaction from a reflex.

Part 4

Acting Out

This part of the book faces some of the most difficult questions about our brains: How do we remember information? What makes one person smarter than another? Where do our emotions come from? By now you have already learned that although we know a great deal about the brain, there is far more that we don't yet know. The areas of memory, emotion, and intelligence are among the most complex in the field of neuroscience, and much of what we discuss in this part of the book remains the subject of great debate and intense investigation.

Chapter 14

Blackboard of the Mind

In This Chapter

- ◆ The long and short of memories
- ◆ Information overload
- ◆ Memory mistakes
- ◆ Why we forget
- ◆ Memory boosters

Unlike the ancient Greeks and Egyptians, who thought the heart was the center of intelligence, modern people know that thinking happens in the brain. Yet most people don't know how the brain controls thinking. They see the whole brain as being a sponge that absorbs information or a computer that stores it, but as you may have figured out by this point in the book, this idea is inaccurate.

Much of the brain has nothing whatsoever to do with the high-level functions of thinking, learning, and remembering. At the same time, no single area of the brain has been identified as the place where these mental processes take place. The cortex is certainly the key area; the prefrontal cortex plays a specific role, as do other substructures in other areas of the brain, such as the hippocampus and the amygdala. This chapter explores where memory fits into the picture.

Memories ...

What is a memory? In general terms, memories are bits of information that have entered your consciousness, been stored in your brain, and now can be retrieved. Memories can be images (photographs or text), smells, tastes, and feelings. They give order to our lives, allowing us to recognize people and communicate with them.

What does a memory look like anatomically? The answer is not exactly clear. A memory may be a circuit or a set of connections between nerve cells. When a person learns new things, the brain creates new circuits that somehow remain active so that the knowledge can be retrieved. A large number of neural connections spread throughout the brain are required for a single memory. Circuits that go unused gradually become disconnected, and the information is lost. That's how a person forgets.

As you've learned in previous chapters, many of the processes of the brain occur automatically, and others involve conscious decisions. Memory can work in both ways. When you are trying to remember something—someone's birthday or what you had for breakfast, for example—and you're aware that you are recalling a memory, your brain is using *explicit memory*.

The unconscious (some experts say *nonconscious* is a more accurate term) type of memory is referred to as *implicit memory*. An example of implicit memory is driving a car. You turn the key and perform all the tasks involved in driving without thinking about how to do them, and you obey the rules of the road (driving on the right side, signaling before turning, stopping when the light turns red) without having to consciously recall them.

Hidden Messages

Those who try to send subliminal messages, such as advertisers, are exploiting people's implicit memories. Subliminal messages take a variety of forms, from naked women disguised as ice cubes in liquor ads (designed to entice you to buy the alcohol) to allegedly phallic symbols in cigarette ads.

IQ Points

The belief in the power of subliminal messages may have begun when advertising promoter James Vicary claimed in 1957 that he dramatically increased the sale of popcorn and Coca-Cola in a New Jersey movie theater by flashing the messages "Drink Coca-Cola" and "Hungry—Eat Popcorn" on the movie screen very briefly. He said that no one could see the images, but they registered in the viewers' subconscious. The story turned out to be a hoax, and the strategy was never shown to work as touted. Now, of course, blatant commercials for popcorn, candy, and soda are shown before the movie.

Subliminal messages are not just aimed at getting people to do things they don't necessarily want to do; they are also a popular technique in the self-help business. Lots of people have been sold on the idea that listening to tapes with subliminal messages can help them improve their memories, lose weight, stop smoking, and produce other positive behavioral changes.

Convinced of the power and omnipresence of subliminal messages, some people believe they have discovered hidden messages in all kinds of media. For example, they play music backward to reveal some supposedly secret (often allegedly satanic) meaning. A few people went so far as to sue the Disney company because they were convinced that several of its animated features contained subliminal sexual messages. For example, a frame in *The Lion King* supposedly has a cloud of dust, grass, and flowers that form the word *sex*.

Because the point of subliminal messages is to reach the unconscious mind, a message can't truly be subliminal if we know it is there. For example, an ad produced by the Republican Party that attacked Al Gore in the 2000 campaign flashed the word *RATS* across the screen for $\frac{1}{30}$ of a second. The word was visible long enough so it wasn't really subliminal, and it did get the attention of Democrats who protested. The Republican Party subsequently pulled the ad off the air.

Apparently, there is no law against subliminal advertising, but the Federal Communications Commission issued an order in 1974 saying that broadcasting subliminal ads was "contrary to the public interest."

Many people are convinced of the effectiveness of subliminal messages, but the evidence suggests that they don't provoke us to purchase products we don't want or help us improve our lives.

Pigeonholes

Memory can be divided into a host of categories. For example, declarative (fact) memory refers to places, dates, names, faces, and events that are stored primarily in the *hippocampus*. The prefrontal cortex is also involved in putting facts in context with the time and space in which they occurred. These facts are relatively easy to forget. We also have memories of skills, such as how to ride a bicycle or tie our shoelaces, which involve a different area of the brain and are much harder to forget than declarative memories. Yet another slightly more abstract type of memory is referred to as sensory memory. This is our knowledge of the world around us at any given time that registers when our senses are initially stimulated.

> **Words of Wisdom**
>
> The **hippocampus,** which resides within the limbic system, is believed to be deeply involved in the storage and processing of information and, specifically, memory and learning.

In addition to categorizing memories in terms of content, we also divide them by their duration; that is, short-term (or working) memory and long-term memory. The following sections describe these two types of memory in more detail.

Gray Matter

Certain types of information are easier to remember than others. For example, most people can memorize images more easily than words. So if you want to teach children what an airplane is, you should show them a picture of one rather than reading them a description of one.

How Short Is Short-Term?

Can you remember what you ate this morning for breakfast? How about what you ate yesterday? A week ago? A month ago? A year ago? You may think that re-calling today's meal is a short-term memory, but technically it isn't. Short-term is not relative to longer periods of time; it is a finite period of about 30 seconds (some researchers define it as a slightly shorter duration, and a few suggest a period as long as a couple of days). The words you just read in the last sentence are now in short-term memory; the words at the top of the page have already left it.

In addition to storing information for only an extremely short time, short-term memory also can hold only a small amount of information. Short-term memory holds about seven chunks of information, give or take two. Can you think of anything that you usually can remember, but only for a few moments, that is this size? Sure you can—a phone number. If you look up a number in the phone book and then close the book, you can probably remember the number long enough to dial it. But you might have trouble if you had to make an international call, because that call would require you to remember more than seven digits.

We know what you're thinking: "But I can remember my phone number for more than 30 seconds!" You remember your own phone number because you've repeated it and writ-ten it enough times to move it from your short-term memory to your long-term memory.

Quiet, Brain at Work

The term "working memory" is sometimes used synonymously with short-term memory. Some researchers, however, distinguish working memory from short-term memory by its immediate use. Working memory can consist of new information, such as the phone num-ber you just looked up, or memorized information, such as the name of your first grade teacher, which you recollect while telling a story about your childhood.

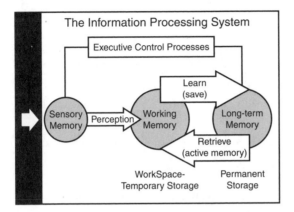

The Information Processing System

Executive Control Processes

Learn
(save)

Sensory
Memory — Perception — Working
Memory — Long-term
Memory

Retrieve
(active memory)

WorkSpace-
Temporary Storage

Permanent
Storage

This diagram gives a sense of how information is processed by the brain. First, it enters the sensory memory, then it becomes part of the working memory. If it is learned, the information may become part of long-term memory. This information can then be retrieved in the future.

This concept may be more familiar to people who understand computers. Suppose you write a letter on your computer. The words on your screen are in the computer's working memory (the cache). If you turn off the computer or exit the program without saving the letter, the information is lost. If, however, you save the document, it is stored on the hard drive of your computer. The hard drive is the computer's long-term memory. When you turn off the computer, the working memory of the letter is gone, but the letter is still in the long-term storage. When you turn on the computer again, the letter is still saved on the hard drive (that is, the long-term memory), and then you recall it to the computer's working memory by opening the file.

Remember Your Childhood?

How long is long-term memory? It can last a lifetime, though most people can't remember events before they were three years old. A variety of factors can influence whether you retain any particular memory. One is the significance of a particular event. Special occasions, such as birthdays, anniversaries, recitals, and athletic achievements, often are especially memorable. You're less likely to recall more mundane events in between. Thus you might remember the first time you rode a bike, went to school, or caught a fish, for example, but you probably don't remember the second or third time you did any of those things. For information to be stored in long-term memory, it must be encoded; otherwise, it is discarded while in short-term memory.

Gray Matter

Researchers believed for some time that long-term memories were accessed through the short-term memory bank. Now, however, it appears that they may be independent.

When information is encoded, it is converted from a stimulus to a form that can be stored in the brain. Codes come in three principal forms:

- **Acoustic coding**   Remembering information based on its sound.
- **Visual coding**   Remembering an image or other visual characteristics.
- **Semantic coding**   Remembering something based on its meaning.

In the preceding section, we suggested that long-term memory is similar to a computer hard drive that allows you to save and retrieve enormous amounts of information. But in Chapter 1 we also argued that your brain is far more complex and extraordinary than any computer; the storage of memory is just another example of that fact. While every few months the size of hard drives increases to the point where they can now hold more than 100 gigabytes of information, your brain can store a nearly infinite amount of data. Some researchers believe the brain also stores the information permanently; others maintain that this storage deteriorates over time.

Long-term memory is sometimes categorized as episodic, semantic, and procedural. Episodic memory is information related to specific events. It can be further broken down into autobiographical memory to describe our personal experiences. Semantic memory refers to factual or conceptual memory. You acquire this information, but you may not remember how or when you acquired it. Examples of this kind of information are, the phone number of your child's school, the alphabet, or the number of inches in a foot. Tasks that you've repeated so often that you can do them without thinking are part of your procedural memory. An example might be tying your shoes or writing your name.

These types of memories can be independent, or they can work together. For example, you can consciously think about the steps involved in tying your shoe (a semantic memory) and perhaps recall the first time you did it yourself (an episodic memory), and then go ahead and quickly tie your shoe without going through the process step by step (using procedural memory).

Time to Decide

When you walk down the street, your brain takes in the information that tells you where the curb is and enables you to identify a crack in the sidewalk, spot a pole, or veer away from an oncoming bicycle. You usually process these individual bits of data immediately, and then typically discard them. For example, you usually don't remember that there was a crack in a particular spot in the sidewalk that you just stepped over. These sensory memories are stored for even less time than short-term memories (less than a second for a visual [iconic] memory and less than four seconds for an auditory [echoic] memory), and then the brain decides whether to store or discard the information. Only something that we need to know for our own basic needs, that will satisfy some personal interest, or that stands out as unusual is typically selected for longer term storage.

Once we've encoded and stored memories, how do we use them? It wouldn't do us much good, after all, if we could save things on the hard drive of our computer and not be able to use them again. Memory retrieval comes in at this point. Once again, this process can be broken down into several categories:

◆ Recall is the process by which you can remember information without being prompted with a part of the memory. This process is exemplified by answering a fill-in-the-blank question on a test.

◆ Recognition occurs when some stimulus from a memory is experienced again, such as when answers are provided in a multiple-choice exam.

◆ Recollection involves reconstructing pieces of memories that go together. This process is used during an essay test.

◆ Relearning occurs when you may have difficulty remembering something you once knew, but you have an easier time after learning it a second time. For example, if you see a picture of the kids from your high school graduating class, you may not remember everyone, but if you see a list of their names at the same time, you'll have an easier time matching the names and faces than someone would who had never known the people before and was just given the photos and list cold.

During the process of remembering, the brain seeks patterns to bring pieces of memories together. The more orderly the pattern is, the higher the probability that you'll remember the information. Suppose we put a collection of randomly assorted square, round, and triangular blocks in front of you. If we then put the blocks in a box and asked you to take them out and recreate what we had done, you would probably have a difficult time remembering the original layout. If, however, we took the exact same blocks and built, say, a house with them, you would have a much better chance of recreating the pattern.

An even more common example relates to memories of sound. Random noise or even a discordant piece of music is far more difficult to remember than a pattern of sounds, such as the sometimes annoying ditties heard in commercials or the more inspirational movements of a classical symphony (think of how readily Beethoven's Fifth Symphony comes to mind after hearing those first few notes).

Code Blue

Amusia is an unusual condition in which people who are able to hear and understand speech and most other sounds cannot recognize melodies. This condition goes beyond the more typical tone deafness many of us have when it comes to singing and is usually a result of damage to the temporal lobe of the brain.

Mozart for Memory

In 1993, researchers at the University of California at Irvine did an experiment in which college students were asked to listen to 10 minutes of silence, a relaxation tape, or a Mozart sonata. Afterward, the students were given a test that showed the ones who had listened to Mozart performed better. Although the effect appeared to be temporary, lasting about 10 to 15 minutes, the researchers argued that music could improve memory because they shared similar pathways in the brain.

This so-called "Mozart Effect" was subsequently tested by a number of other researchers in different settings with different subjects and types of music and sounds. The original findings could not be duplicated, and now the notion that listening to a particular sound or type of music can enhance memory is considered dubious. Nevertheless, a great many people still believe that memory and general intelligence can be influenced by music, and parents continue to play Mozart for babies in the womb because they're convinced it will help their child develop.

Fill in the Gap

Chapter 10 explained how optical illusions are created by the brain's effort to interpret images to fit known patterns that don't match the actual image. The brain engages in a similar process when it is working with other types of memories. People organize information in a way that makes sense to them, and this system of organization sometimes results in a memory that is inaccurate.

In one experiment that illustrates this phenomenon, people were given a list of words that included sour, candy, sugar, bitter, good, taste, tooth, knife, honey, photo, chocolate, heart, cake, tart, and pie. They were then asked a series of questions about whether a particular word was on the list. When asked if the word *sweet* was on the list, a large number of people said "yes." If you look at the list again, you can easily see the logical association, but sweet is not on the list. You can look at this result as an example of people having poor memories, but that's not the way scientists interpret it. They suggest that this result shows that we learn by relating new information to existing memories.

False Memories

Of course, not everything that we remember is accurate. Sometimes we get facts and figures wrong. We also can forget incidents that happen to us or remember them differently than the way they actually occurred. These inaccurate recollections are called *false memories*.

In most cases, we expect this imperfection of memory. Though it's annoying, it isn't serious. However, serious problems, not to mention litigation, have been caused by instances where mental health professionals have allegedly created false memories in their patients through their therapeutic techniques.

How can this happen?

Creating false memories is easier to do than you might think. First, most people expect to remember the past and may feel guilty if we don't. Suggestions that people should remember something they don't provide an incentive for people to try to come up with a memory. Memory also can be influenced with hints, prodding, or false accusations. For example, if someone very convincingly says he saw you do something you don't remember doing and did not do, he may be able to convince you that you did do it. And if that's not bad enough, you are liable to begin to invent details about the behavior that never happened.

Many times these memory mistakes can occur quite innocently. For example, in the course of researching a book on World War II, former prisoners of war (POWs) would understandably have difficulty remembering incidents that happened to them 50 years ago, but if a fellow POW had a particular memory, even if it was inaccurate, other POWs would often confirm the false account.

Gray Matter

In one case of false memory, a church counselor helped a woman remember during therapy that her clergyman father had raped her and her mother had helped hold her down. One of her memories was that she had become impregnated and had to have an abortion. The sensational story made the papers, and the woman's father was forced to resign his post in disgrace. The woman was later found to be a virgin and had not had an abortion. She sued the therapist and received a million-dollar settlement.

The controversial cases that have gone to court have involved charges that therapists made suggestions that led patients to recall false memories that were very damaging, such as the belief that they were molested by a caregiver as a child.

If a person's memories can be independently corroborated, then true memories can be distinguished from those that are created. However, without such corroboration, demonstrating that memories that have been recalled through suggestion are not authentic long-buried memories is difficult, if not impossible.

Magnetic resonance imaging indicates that the same areas of the brain are active regardless of the validity of the memory. This knowledge leads researchers to surmise that false memories are stored in the same area of the brain as true ones, which may explain why someone can become convinced of something that did not happen.

Drawing a Blank

You've probably never met someone with amnesia, but you've undoubtedly encountered the condition in the movies. The fictional case usually involves someone who has forgotten everything, but amnesia is often partial and may involve the loss of either old or new memories. The effect can be permanent or temporary.

Have you ever wondered how it could be that someone with amnesia in the movies can't remember anything, but still speaks perfectly and knows how to live an otherwise normal life? If you really lost your memory, wouldn't you forget how to speak and how to drive a car and buy groceries? Obviously, a movie would be difficult to follow if the main character suddenly couldn't function, but this is more than a mere plot device. Amnesiacs don't lose the ability to communicate and are still aware of social customs; for example, they don't suddenly forget that they are supposed to wear clothes.

IQ Points

The Internet Movie Database lists 88 films with amnesia as part of the plot. In 2001, for example, Jim Carrey played a character who loses his memory in the film, *The Majestic*. An even better movie (fascinatingly told in reverse order) is *Memento*, which features Guy Pearce as a man who suffers a head injury and has no short-term memory. He then has to tattoo notes on his body and take photographs to remind himself of new things that happen to him. This character suffers from what's called anterograde amnesia, the scientific term for loss of short-term memory.

Occasionally, amnesia is a product of a severe psychiatric illness or extreme stress. This "hysterical amnesia" may occur suddenly, and sufferers find themselves in strange places unable to remember their names or other details about their lives. The onset is often a feeling of being overwhelmed by grief or shame or some other powerful emotion that makes it difficult or impossible for the person to face reality. This type of amnesia often can be treated with therapy.

Usually, amnesia is a result of an injury to the brain. Damage to the limbic system, in particular, affects the storage of memories. Depending on where the injury occurs, different types of amnesia may result. For example, damage to the visual cortex in the occipital lobe may lead to visual amnesia, which is the inability to recognize printed text or objects. An injury to the temporal lobe can produce auditory amnesia, which is inability to remember words. The inability to identify objects by touch, called tactile amnesia, can be caused by injury to the cortex in the parietal lobe. Frequently when the injury heals or when a tumor that has affected memory is removed, the amnesia goes away.

Alcoholism is another cause of amnesia. Heavy drinkers may experience blackouts and find themselves unable to recall what happened when they were intoxicated. Prolonged abuse of alcohol can damage brain cells and cause permanent damage to short-term memory and less extensive damage to long-term memory. This kind of damage is called Korsakoff's syndrome and is thought to be caused by a thiamine deficiency. It affects the thalamus with degenerative changes.

In the movies, the amnesiac usually forgets who he or she is and everything about his or her past, but has no problem with new memories. In reality, people with amnesia have more trouble remembering events that occur after their injuries, which is known as *antero-grade amnesia*. *Retrograde amnesia*, the type where a person loses the memory of events that occurred before the illness or head injury, is rarer and usually is accompanied by antero-grade amnesia.

After a head injury, short-term retrograde amnesia is common. For example, the injured person does not remember the events that immediately preceded the accident or blow to the head, and the memory of these events may be lost forever. The person may recall some of the events leading up to the injury, or the injury may be such that the person doesn't remember anything. Memories of events closest to the time of the injury are the least likely to return.

Going, Going, Gone

You are not supposed to remember everything. Forgetting is part of our biological makeup, and it is to our benefit. Think about it for a moment. Sure there are lots of things you wish you could always remember, such as the names of all the U.S. presidents, the sight of the sun setting over the ocean in Hawaii, and the taste of the first time you ate chocolate. But there are also lots of unpleasant memories that you are better off not remembering, such as the pain of a broken bone, the grief over the death of a loved one, or the taste of spoiled milk. Total recall would be physically and emotionally overwhelming.

A number of theories have been advanced to explain why we forget. One is that memories just fade away from disuse. A common analogy is a path in the forest that eventually becomes overgrown, erodes, and disappears if it is not used and tended. Another idea, which is called *interference theory*, is that we need to erase old memories because they somehow interfere with storing new information, or conversely, acquiring new memories inhibits remembering old information. Freud argued that memories aren't necessarily forgotten; they may instead be repressed so they are unconsciously blocked from our awareness to protect us.

Age is also a factor in forgetting. Brain cells begin to die in early adulthood and continue to do so throughout a lifetime. Recent research suggests the situation may not be as bad as once thought because some neurons may be regenerated in the hippocampus.

Loss of brain cells is not the only reason for deteriorating mental functions as most people age. Another reason is believed to be a decrease in blood flow to the brain. This change is not uniform; instead the reduction of blood flow is greater in areas of the brain responsible for concentration, spontaneous alertness, and the encoding of new information.

CAUTION

Code Blue

Trauma to the head can also cause mild to severe memory loss. This loss may occur to football players who take a lot of blows to the head and, especially, to boxers, who can become "punch-drunk" after too many fights.

In the extreme case of Alzheimer's disease, which we discuss in detail in Chapter 18, the sufferer lives in a state of confusion and loses both short and especially long-term memory to the point where they cannot recognize their own families. In every disease where dementia is prominent and where cognitive functions are lost, loss of memory is also common.

On a more positive note, many people function quite well as they age, and some excel. To name just two, Stanley Kuntz was named poet laureate of the United States at age 95, and Grandma Moses didn't start painting until she was 76.

Bulking up Memory

Some people are born with a better memory than others. You've probably heard of someone having a photographic memory, which is the ability to look at something once and not only instantly recall it, but remember the information indefinitely. This talent, incidentally, doesn't necessarily mean someone is especially intelligent. For example, the mathematical ratio pi (the ratio of the circumference to the diameter of a circle) begins 3.1415 and continues indefinitely with no pattern. A man known as Rajan memorized the first 32,000 digits, and could perform other amazing memory feats associated with pi, but did not have an exceptional score on intelligence tests.

No one can explain why only a few people are blessed with this ability, but the good news is that we can all improve the memories that we do have using the following techniques.

The simplest way for most people to remember a specific piece of information is to repeat it. If you go over and over the same list or task, you can increase the chance that you'll learn it. Different people may respond better to different types of repetition. For example, if you tend to remember better when you hear material, you might do better to repeat it out loud. If you are a more visual person, pictures can be useful tools. For other people, the process of writing and rewriting the same material works best. Good old reliable flash cards still are an effective tool for memorizing spelling words, multiplication tables, and other information that does not require analysis.

A mnemonic device is another proven way to remember lists of information. This device usually consists of a string of words or an acronym that you can associate with the information you need to learn. Here's one from Chapter 8, **O**n **O**ld **O**lympus **T**owering **T**op **A** **F**inn **A**nd **G**erman **V**iewed **A** **H**op. Each of the bold letters represents the first letter of one of the 12 cranial nerves. Unless you're going to medical school, you probably won't encounter that one, but many of us learned a mnemonic in school to help us remember the names of the Great Lakes. The first test is whether you remember the mnemonic, and the second is whether you still can use it to recall the lakes. You'll find the answer at the end of the chapter. (No fair peeking!)

A similar approach is to make up a song or rhyme that reminds you of what you are trying to remember. In school, how were you taught to remember the year Columbus discovered America? Remember, "In fourteen hundred and ninety-two, Columbus sailed the ocean blue"?

One way to remember anything is to associate it with something that sticks in your mind. This practice is referred to as the *method of loci* (or places). For example, if you are trying to memorize a shopping list, you can imagine all of the items in different parts of your house. Think of a carton of milk sitting on the kitchen table, a loaf of bread nailed to the wall, a can of tomato sauce in your bed. When you get to the store, picture yourself going through the house. This technique may sound silly, but you will probably remember the list. In fact, the more bizarre the associations are, the better this technique usually works.

Words of Wisdom

The **method of loci** dates back to ancient Greece when Greek orators would use this method to help them memorize speeches. One story relates that the Greek orator Simonides of Ceos was at a banquet to give a speech and stepped outside for a moment. At that moment, the building collapsed, and everyone inside was killed, and their bodies were mangled beyond recognition. Simonides identified the bodies of the guests based on where he remembered them sitting or standing before he left the building.

Another effective technique is to group large sets of information into smaller chunks. This technique works especially well if you need to remember a list of numbers. You probably already do this with your burglar alarm, ATM identification numbers, and other codes. Are any of your codes combinations of birth dates, anniversaries, or other significant moments in your life? If so, why did you choose them? You chose them because they were easy to remember, of course.

Code Blue

Security experts advise against using familiar dates and numbers, such as birthdays and anniversaries, for passwords because they are the first things that criminals wanting to steal your password are likely to try. It is better to use a combination of numbers and letters that aren't related to common events in your life.

Let's say you didn't get to choose your own code and needed to remember one given to you for accessing an Internet site. The code is 361967401BG. We would break the code up this way:

36 1968 401 BG

We could remember those four chunks by thinking of a relative's birth date (1936), the year of the Mexico Olympics (1968), a retirement plan (401), and the first prime minister of Israel's initials (BG for Ben-Gurion). Then, instead of trying to remember the numbers, we could think of our relative watching the Olympics with a pension for Ben-Gurion.

Lifestyle changes can also improve memory. First, get lots of sleep. Fatigue is an enemy of memory. Pulling an "all-nighter" to cram for an exam may be necessary and may seem productive, but the lack of sleep may also make it harder to recall the information you studied and to concentrate on the test.

Stress can also affect memory. Some people perform better under pressure, but most of us don't. Lots of people can win prizes and money playing game shows from the comfort of their living rooms, but once they get under the hot lights in front of a TV camera that is beaming their face out to millions of viewers, the questions can suddenly seem a lot more difficult. When you relax, it is usually easier to focus and to draw on your memory.

Eat Smart

Did your parents ever feed you "brain food?" They believed, as most people do today, that certain foods can improve your memory and intelligence. Here are a few examples:

♦ Sugar is important because it supplies glucose to the brain and this is the fuel that makes the brain work.

♦ Protein foods, which can be found in pizza, nuts, eggs, and meat, provide amino acids that are important building blocks for brain tissue. Amino acid also can be found in peas and rice.

♦ Vitamin B complex is related to memory and learning.

♦ Vitamin B1 (thiamin), found in oatmeal and green peas, is needed to process sugar, and this helps maintain healthy nerves and aids memory.

♦ Vitamin B3 (niacin), found in turkey and tuna, is important for red blood cells that carry oxygen to the brain.

♦ Vitamin B12, found in meat, eggs, and dairy products, is important for balance.

◆ Vitamin C, found in broccoli, cauliflower and citrus fruits, aids in the production of dopamine.

◆ Chocolate may increase serotonin levels in the brain, which are important as mood elevators.

◆ Vitamin E, found in cereals, fruits, poultry, nuts, and seafood, is best known for increasing sexual potency, but some researchers believe it may also improve memory.

Excessive intake of any one or all of these vitamins may cause severe problems and may make worse the things you're trying to improve. Various herbs and supplements are also said to be especially beneficial, but there is some controversy regarding these benefits.

Foods can also have a harmful effect on the brain that is often overlooked in the diagnosis of illness. For example, food allergies to coffee (probably to the caffeine), milk, chocolate, sugar, and wheat may cause psychological problems such as anxiety or depression. It is important to understand this connection because recognition that an allergy is present may explain the mental disorder. Someone who is depressed and is found to be allergic to dairy products, for instance, may be cured of the depression when they stop drinking milk.

In addition, keeping mentally active by reading, writing, doing crossword puzzles, and even walking has been shown to be helpful in preventing memory loss.

Try to Remember ...

Storing information in your brain is the easy part. The tough thing is getting it back out when you need it. Going back to the computer analogy, think about the steps required to find a particular file on your computer. You can run a search or look through directories. The bigger the disk drive, the more difficult it becomes to find the particular file you want. Multiply that problem by an exponential factor, and you get a sense of how difficult it is to retrieve a particular fact from the vast amount of knowledge you're carrying around in your head.

Gray Matter

Have you ever tried to remember something and have the feeling that it's right there at the tip of your tongue, but you can't quite get it out? You're not alone. This common phenomenon has been studied for years. No one knows why it happens, but apparently the process of retrieving a long-term memory breaks down in what seems to us like the moment before recognition. One theory is that the brain calls up an inaccurate piece of information that blocks the correct memory from emerging. Another suggestion is that you don't have enough information for the brain to determine what else needs to be retrieved.

We all have plenty of moments when we can't come up with information we know, such as the word for the gate of a castle (the portcullis), the name of the last Soviet president (Gorbachev), or the ingredients of a Caesar salad (anchovies, olive oil, lemon, egg yolk, Worcestershire sauce). What is remarkable is how quickly and easily we can retrieve most information. We're able to do this because of the way the brain organizes and cross-references our memories.

(By the way, the mnemonic for the Great Lakes is HOMES, which stands for Huron, Ontario, Michigan, Erie, and Superior. Did you remember this?)

The Least You Need to Know

- A memory is thought to be a neuronal circuit, which deteriorates if unused and leads to forgetting.
- Small chunks of information can be retained in short-term memory, but information must be encoded to become a lasting long-term memory.
- Remembering is a process that requires the brain to put pieces of memories together in a meaningful way.
- You can improve your memory through repetition and tricks such as mnemonic devices, songs, and the method of loci.

Chapter 15

The Feeling Brain

In This Chapter

- ◆ Are you crying because you're sad or sad because you're crying?
- ◆ How are thinking and feeling related?
- ◆ What can you tell about emotions by looking at someone's face?
- ◆ How are emotions related to stress?
- ◆ Is a high EQ better than a high IQ?

What is love? Happiness? Hate?

They are emotions. But what are emotions? Put simply, they are feelings and sensations. Emotions have a strictly physiological component, but they may have a cognitive component as well, one that is not easily visible, measurable, or definable. One question is what comes first: the response of the body or the thoughts that we call emotion?

Describing emotion is difficult, and determining the brain's role in feelings and behaviors is even more problematic. Separating the biological from the environmental influences is even harder. Nevertheless, this chapter takes up the challenge.

Feelings First?

Most of us have felt anxious at one time or another. When you're nervous or under stress, you may feel your stomach tighten as though it's in a knot. Here's the tough question, though: Does your stomach tighten up because you're anxious or do you feel anxious because your stomach tightens?

For many years, the conventional view was that a stimulus produces an emotion, and the emotion determines a behavior. If you see a big dog barking and running toward you, you get scared, and then you run away. But at the end of the nineteenth century, William James turned this view on its head by suggesting that the stimulus causes a physical response in the body, and the physical response then determines the emotion. Thus, for example, James believed that when you see the big dog, your body's fight-or-flight mechanism becomes active and *then* you feel scared instead of feeling scared first.

Working with Carl Lange, James further refined his theory by arguing that a stimulus, such as seeing a dog approaching, triggers impulses in the sensory cortex. The sensory cortex, in turn, signals the autonomic system, which raises the level of alertness of the body. This heightened sensitivity feeds back additional signals to the brain that generate an emotion.

Walter Cannon attacked this theory in the 1920s. He argued that the James-Lange explanation didn't take into account the speed with which emotions are generated. Cannon argued that the process of sending messages via the autonomic system took too long to account for human reactions. He also maintained that people still said they felt emotion even if their sympathetic nervous systems were detached. Furthermore, visceral reactions frequently are common to multiple emotions. For example, your stomach may feel like it's in a knot if you are very anxious about having to speak in public or because you are angry about something someone has said to you or because you are sad about an unexpected death in the family.

Cannon, along with his colleague Philip Bard (no relation to the current author), proposed an alternative theory that says emotion originates in the thalamus. This region of the brain transmits part of the incoming sensory message to the cortex, which registers fear, anger, joy, or another emotion in the conscious mind, and simultaneously sends another part of the message to the hypothalamus, which stimulates the physiological response (remember the hypothalamus controls autonomic functions). Put another way, Cannon and Bard believe we feel emotion in our heads first, and then our bodies react (for example, by tightening the stomach muscles, sweating, and increasing heart rate). On the other hand, James and Lange hold that feelings begin with the physical reaction and then the brain registers the emotion.

Gray Matter

The stimuli that cause emotion can be internal as well as external. For example, an external cause of anger might be someone hitting you. If you were ill and the sickness caused severe pain, that internal stimulus might also anger you.

Other research suggests that some sensory inputs travel to the thalamus and the *amygdala* but are never passed on to the cortex. Consequently, this information is not fully processed and evaluated and may produce a rapid, "unthinking" reaction. Other emotions, however, result from a more thorough processing of information that has been passed from the thalamus to the cortex.

Still another idea is that the visceral response tells you that it is appropriate to feel an emotion, but it does not determine the type of emotion. According to psychologists Stanley Schachter and Jerome Singer, the higher order processes are then required to decipher the appropriate emotion from the situation in which the visceral response occurs. For example, both a Doberman chasing you down the street and a competitor in a marathon you are trying to outrun will stimulate similar physiological reactions, such as increasing your heart rate. The emotion you attach to this physical response is related to the circumstances, so in these examples the scary dog prompts a feeling of danger and a sense of fear, while the other runner provokes excitement. This cognitive theory explains why different emotions sometimes have similar visceral responses. For example, tears may flow from joy or sadness depending on whether you are at a funeral or a wedding.

Code Blue

The study of emotion is problematic. One difficulty is that scientists study outward behavior, which may not be the same as a person's internal feelings. Another troubling issue is that most research must be conducted on animals because parts of the brain must be destroyed to determine the areas that influence specific emotions. Given the greater complexity of the human brain, these animal studies may not accurately reflect the relationship between brain areas and emotional responses. For example, humans have a more highly developed forebrain than animals do, and this forebrain has greater control over the functions of the limbic system.

Although many scientists believe that emotions are determined by biological factors, others take an entirely different approach and argue that genetic factors are not important or at least not as important as experience and social learning. The cognitive and constructive theorists believe that emotions are closely related to thought processes. Self-awareness, for example, allows us to identify the emotions we feel. Psychologist Magda Arnold and others have argued that stimuli are evaluated before eliciting an emotion.

To go back to the example of the barking dog, cognitive theorists would suggest that you associate big barking dogs with danger, and that association produces the emotion of fear. The body's physical reaction then moderates the intensity of the emotion so that the increase in heart rate and other changes associated with heightened alertness enhance the level of fear.

One problem with these theories is that unless cognition takes into account an extremely broad range of mental activities, it cannot explain the expression of certain emotions that appear to have no relation to conscious thought. For example, infants show expressions to indicate anger and happiness that are unlikely to be associated with the ability to evaluate the pain that produced the anger or the parent's action that prompted the baby's smile. In addition, we also sometimes act on the basis of emotions that are contrary to rational decision-making. These gut-level emotions include feelings that we should do something (such as buy a lottery ticket because we have a hunch we've got a winning number) even though our minds are telling us we shouldn't (because the chance of winning the lottery is remote).

As you might imagine, the study of emotion is a complex and controversial one and includes many more theories than those that are presented in the following sections.

A Range of Emotion

How many emotions do we have? If you think hard about it, you can probably come up with a lengthy list. Paul Ekman reduced the number to six primary emotions: happiness, disgust, fear, surprise, anger, and sadness. Other emotions that come to mind are most likely some variation of these six, differentiated mainly by the intensity of feeling. For example, happiness would include love, euphoria, and optimism.

Robert Plutchik came up with eight emotions and suggested that experience is a blend of more and less intense emotions. For example, he sees love as a mixture of joy and trust. If you combine joy with anticipation, however, you get optimism. Similarly, combining disgust and anger produces contempt while anger mixed with anticipation results in aggressiveness. Isn't it interesting that according to this theory anticipation can lead to such radically different emotions depending on what else you are feeling?

The number of emotions is less important for the purposes of this book because we are more concerned with the brain's role in producing emotions than the precise definitions or descriptions of emotions. What researchers are learning is that the more primitive, instinctive survival feelings, such as anger, are rooted in the limbic system, but they are moderated by the critical thinking performed in the cortex. Perhaps this is why there appears to be an inverse relationship between the activity in the cortex and the amygdala; that is, when the cortex is especially active, the amygdala is relatively quiet and vice versa.

> ### Words of Wisdom
>
> In 1977, Roger Brown and James Kulick suggested that people could so vividly recall what they were doing at the time of John F. Kennedy's assassination because shocking events such as these activate a special brain mechanism they referred to as **Now Print.** The Now Print mechanism acts like a camera's flashbulb, freezing in our mind whatever happens at the moment when we learn of the shocking event. These memories are referred to as "flashbulb memories."

The amygdala also appears to play a role in memory by becoming active when emotion is involved in the memory. For example, a memory of how to solve a geometry problem probably has no emotional association for most of us who are not mathematicians and is encoded in the hippocampus. If, however, we see a tragic incident, such as the bombing of the World Trade Center, the memory may trigger the amygdala, which is attached to the hippocampus, and add an emotional component to the memory. Note that the bombing itself is simply a sensory input. Something about it, however, registers with the part of our brains that infuses it with emotional content and makes us have a feeling about it—in the World Trade Center case, probably sorrow and anger—that distinguishes this memory from others.

Looking the Part

We often wear our emotions on our faces. This relationship between emotions and facial expressions implies that signals associated with specific feelings are sent to particular facial muscles, and the resulting looks convey to others a sense of what we are feeling. These facial expressions are also universal; that is, people from different cultures around the world have similar expressions for the same emotions, and each understands these expressions to mean the same thing.

This relationship is not exact. A smile, for example, can mean happiness or simply be an outward effort to conceal a different feeling. Think how often you smile at someone even if you are feeling badly. If you don't know the context of the outward appearance, determining which emotion is being expressed is sometimes difficult. For example, determining whether someone who is crying is happy or sad usually requires more information than just the presence of tears.

These few examples demonstrate the relationship between emotion and the physical expressions the brain calls for:

♦ **Anger:** Pursing of the lips, flaring of the nostrils, and narrowing of the eyes

♦ **Happiness:** A smile, wrinkles around the nose, and wide, bright eyes

♦ **Sadness:** Trembling of the lower lip, drooping of the head, and a frown

Chapter 8 explained that two cranial nerves are involved in facial expressions. The facial nerve (VII) sends signals from the brain to the facial muscles while the trigeminal nerve (V) provides sensory information from those muscles to the brain. No one knows how, but some scientists theorize that this feedback loop is involved in emotion.

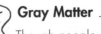

Gray Matter

Though people around the world may have similar expressions for an emotion such as fear, they may not be afraid of the same things or react to fear in the same way.

Going Ballistic

Given that scientists must conduct their experiments on animals, they have focused on emotions that are relatively easy to define and observe, such as anger and fear. The complexity of the relationship between the brain and emotion is exemplified by the finding in some studies that destroying the amygdala has a taming effect on animals while other researchers have found the opposite is true, that animals become more aggressive. Trying to make sense of these results, scientists have suggested that different areas of the amygdala affect the hypothalamus in such a way that one or the other behavior is produced.

Animal experiments in the early twentieth century demonstrated that damage to the frontal cortex had an impact on the behavior of chimpanzees, which prompted Egas Moniz to adopt the idea of removing part of the brain to treat emotional disorders, particularly people who were especially aggressive. The operation, in which the connections between the frontal lobes were severed from the rest of the brain, became known as a prefrontal lobotomy.

Given what we now know about the way functions are spread throughout the brain, it shouldn't be surprising that lobotomy patients suffered numerous side effects. For example, they no longer exhibited extreme rage, but they had other types of emotional overreactions, such as unusual feelings of euphoria and a dampening of social inhibitions. Because of the severity of these side effects, lobotomies are no longer performed.

Oh Happy Day

Though it may sound like something out of a porno movie, your brain has a pleasure center, or more accurately, centers. During neurosurgery, doctors have found that stimulating certain areas of the brain, including the temporal lobes and the hypothalamus, produces feelings of pleasure, optimism, euphoria, and happiness.

Gray Matter

Have you ever noticed how different odors can affect your emotions? A sweet scent may give you a pleasurable sensation, but a pungent odor may irritate or disgust you. This reaction may be explained by the fact that the impulses from the olfactory system travel to the amygdala as well as the olfactory cortex.

This finding does not mean that emotions such as happiness are solely a product of such stimulation. Complex emotions are likely to involve a variety of brain areas. For example, think of some of the things that make you happy. Any memory of a person, a sensation, or an activity requires the activation of the cortex. The higher areas of the brain put together the pieces of an experience such as a barefoot walk on a beach under a moonlit sky with someone you love, and the memory produces a feeling of happiness.

Scientists are trying to study this process through various experiments in which brain scans are conducted while people do things that are pleasurable. Early results suggest that positive emotions stimulate certain areas of the brain and inhibit others. For example, people who are happy have more activity in their left prefrontal cortex, but activity in the amygdala is inhibited. People who are unhappy or depressed appear to have greater activity in the amygdala.

Think Positive

You've probably noticed that when you are in a good mood you act differently than when you are in a bad one. Research shows that a positive outlook can have a number of beneficial effects on behavior. For example, happy people tend to interpret their environment in more positive ways, think more creatively, have greater feelings of empathy toward others, and even have better memories. They also are more open to information, feel better about what they are doing or thinking, and tend to be more open to learning.

One theory to explain these effects is that the information stored in memories has positive and negative qualities and that emotions can affect which qualities are stimulated.

On the other side of the coin, researchers have suggested that people with negative outlooks have greater activity in the ventromedial prefrontal cortex. A recent study by scientists at the University of Minnesota, Vanderbilt, and the Veterans Affairs Medical Center in Minneapolis suggests that this part of brain acts as a volume knob to control emotions and that people who are affected in this way have a higher risk of depression and anxiety. The study also found that lower activity in this region of the cortex does not translate into a more positive or happier outlook.

Taking Control

Some aspects of emotion appear beyond our control. We know, for example, that neurochemicals in the brain can affect emotion. As we'll discuss later, scientists believe that depression is related to chemical imbalances, and studies prove that medication can help correct these imbalances.

We also can take conscious actions to control emotion. Rewards and punishments can affect our behavior. If a child is angry, he or she can be taught how to handle that anger. Society also imposes rules that shape behavior. Angry adults may be tempted to strike the people who have angered them, but they know that this action is against the law and can get them in trouble, so they may control their reactions to the emotion.

IQ Points

Charles Darwin's theory of evolution has also been influential in the study of emotion. His observations led him to believe that the emotional behavior of animals was one of the aspects of their adaptive development.

It may also be possible to change emotions by altering the way they are interpreted. For example, if people feel that it is shameful not to respond to someone who is threatening, they may be motivated to hit the threatening person. If, however, the people learn that there is no shame in walking away, they can do so without feeling badly. This kind of control involves conscious decision-making involving the cortex.

The Prickly Things on Your Neck

The complexity of the processing of emotion is amazing. Think about the reaction to the sound of a car horn. The incoming sound produces nerve impulses that are relayed to the auditory cortex. The auditory cortex analyzes this information and sends a message to the amygdala indicating a possible threat of an onrushing car. The amygdala stimulates the hypothalamus, which triggers the autonomic responses to alert the body with increased blood pressure, faster heartbeat, and sweat, and the emotion felt is fear.

The nearly instantaneous reaction, which prepares the body for an emergency, is an important defense mechanism that helps protect you from harm. This protection might not be possible if you had to think first and analyze the threat. As the autonomic responses are kicking in, the cortex is already evaluating the stimuli and telling you either that the horn is from an onrushing car that is heading toward you or that it is beeping at some other car that hasn't seen the light turn green. Because of this strong reaction, and the activation of the hippocampus and the cortex, the memory of this incident is likely to be stronger than one that is not accompanied by the dramatic physiological changes. The differences in emotional reactions of different people to the same events may be due to varying sensitivities of the amygdala.

Specific areas of the brain also have been shown to have a direct impact on fear. Stimulating regions of the temporal lobes produces an intense feeling of fear or dread. In other experiments, removal of the amygdala has eliminated fear; for example, such a procedure can make rats lose their fear of cats.

CAUTION **Code Blue** _____

Fear may be the most powerful emotion. Even when you know, logically, that there is no reason to be afraid, you may not be able to help yourself. Charles Darwin tested his ability to overpower fear when he went to the zoo and stood in front of a snake. Thick glass separated them, and Darwin put his face against it, determined not to react if the snake struck at him. When the snake did strike, he leaped backward. Darwin was surprised that his will and reason could not overcome his fear.

Life-Changing Stress

We all have coping mechanisms that allow us to handle different levels of stress, but everyone has a breaking point. Research suggests the sensitivity of the amygdala determines this threshold. Everyone has traumatic experiences that rattle their emotions and create stress, but some experiences are so severe they can provoke extreme levels of anxiety, cause nightmares, and stimulate flashbacks.

People who have these reactions may suffer from post-traumatic stress disorder (PTSD). In PTSD, remembering the traumatic experience or having a similar one can trigger the same emotional and physiological response as the one the person experienced with the initial incident. We might think of the stereotypical scene in a movie about veterans of the war in Vietnam who are tormented by memories of horrible incidents during the fighting. The more often this happens, the more deeply ingrained the memory becomes, and the accompanying reactions grow more intense, stimulating flashbacks and nightmares that can ultimately become debilitating.

Gray Matter

Israeli researchers found that mice under stress produced an abnormal form of the brain protein acetylcholinesterase, which interferes with the neuron circuits. The mice's neurons became hypersensitive, and their brains had unusually high levels of electrical activity for some time after the experiment, suggesting that traumatic experiences and repeated stress may affect brain chemistry. This research is just one of many new efforts to explain PTSD.

Lie Detector

We've all seen movies and television programs where criminal suspects are hooked up to polygraph machines to try to determine whether they are lying. The test is based on the assumption that liars are anxious and that this emotion produces measurable changes in skin and blood pressure. As these shows always point out, such tests are not foolproof. Some people can lie without displaying any anxiety or can be trained to "beat" the polygraph. On the other hand, innocent people can sometimes become so nervous or be so intimidated that they become anxious and fail the tests.

Scientists believe a more reliable test for lying may be a scan that could detect changes in the brain when people do not tell the truth. In one experiment, volunteers were given ordinary playing cards and $20. They were then placed in an MRI. A computer would generate different playing cards, and the volunteers were told to lie about whether it was the card they were holding. The subjects were encouraged to be deceptive by being told that they would get more money if they could fool the computer into believing them. The experiment was set up so the researchers always knew whether the volunteers were fibbing, and the researchers could therefore study the MRI results to see whether any noticeable differences occurred when the volunteers were lying. The results indicated that increased activity was detectable in several areas of the brain when lies were told. In the future, such tests may eventually have applications in criminal courts.

Other scientific advances may help to predict behavior, but, as is the case regarding the issue of cloning, ethical questions are raised as to whether we should do things just because we can. For example, what if it were possible to find a biological basis for criminal activity? Should people who have brains with these characteristics be treated in a special way in an effort to prevent them from becoming criminals in the future?

Sound far-fetched?

We've been making policy on this basis for a very long time. During the early twentieth century, for example, many states passed laws that allowed compulsory sterilization of the mentally retarded, the insane, and the habitually criminal because of the belief that they could pass these traits on to their children. Based on what was known at the time, the laws seemed justified, but the scientific conclusions were wrong. Given the history of making decisions based on scientific knowledge that later turned out to be incorrect, many people are uncomfortable with the notion of trusting scientific theories to make predictions about behavior.

Emotional Intelligence

In Chapter 16, we discuss multiple intelligences and the ideas of Howard Gardner. He suggested that people have interpersonal and intrapersonal intelligence that relate to how people interact with others and behave on their own. These two categories come under what other researchers call emotional intelligence. The emotional quotient (EQ) reflects a person's abilities in five general categories:

- **Self-awareness:** Knowing your own feelings
- **Managing emotions:** Knowing how to handle your feelings and understanding what's behind them
- **Motivating oneself:** Exercising emotional self-control, channeling your feelings in the right direction, and delaying gratification and stifling impulses
- **Empathy:** Sensitivity to others' feelings and appreciating their point of view
- **Handling relationships:** Having good social skills so others have a positive feeling about you

No direct relationship exists between IQ and EQ. Some people are emotionally and intellectually "smart," and others are not. Some have more of one than the other. Most people place a high premium on IQ, but the EQ may be no less important in determining how successful a person will be in work or social situations.

Why are some people more compassionate toward others? Is it a learned behavior or one that is somehow hard-wired in the brain? The secrets of EQ are at least as great as the mysteries surrounding IQ.

The Least You Need to Know

◆ Emotions are feelings and sensations that may be produced by a combination of physiological and cognitive processes.

◆ The cortex is involved in thoughts about our feelings, but the emotions themselves also appear to be rooted in the limbic system, especially the amygdala.

◆ Stimulating certain parts of the brain can produce specific emotions, such as fear and rage, which suggests that the brain has centers for certain feelings.

◆ The way people understand and handle their emotions and relate to others has been described as emotional intelligence, and a person's emotional quotient (EQ) may be as important as his or her IQ in determining that person's ability to function in society.

Chapter 16

You've Got Brain

In This Chapter

◆ Multiple intelligence
◆ Wise guy (and gal)
◆ What's your IQ?
◆ Prodigies of all kinds
◆ Using your whole brain

What do you call someone you consider smart?

When you were a kid you probably called them a "brain." Why? Because you knew early on that the seat of intelligence was in the brain. It makes you wonder what the ancients (who believed that intelligence resided in the heart) used to call people they thought were bright. "Hey, that Aristotle is a real heart."

What Is Intelligence?

From the time we are infants, people attempt to evaluate our intelligence. Parents are elated when their babies say their first words and show any signs of knowledge. If children read, write, or do arithmetic at an early age, we begin to think of them as being especially smart. And if they can play music with any facility, thoughts of prodigies like Mozart pop into our heads.

IQ Points

An academic journal once asked 14 prominent psychologists for a definition of intelligence and received 14 different answers.

How quickly you learn those initial skills, however, doesn't necessarily signal the level of your intelligence. Infants who learn fast don't appear to have any greater intelligence than those who get a slower start. The child who reads at age three won't necessarily be a better reader, nor have more intelligence, than the one who starts reading at age five.

In the early years of development, intelligence is measured primarily by parental expectations and comparisons with what other children seem capable of doing. Once children enter school, however, their intelligence is tested, measured, and compared in a variety of ways. Standardized exams are used to rank children with all others who have taken the test. Teachers assign grades to evaluate their work. The grades themselves may not be a true measurement of intelligence, but instead may reflect a specific performance on a given task, the child's motivation to achieve, how much time he or she devoted to studying, and other variables that may be unrelated to intelligence. Still, teachers use the grade to assess intelligence, and a student's peers make their own independent evaluations based not only on grades, but also on perceptions of whether someone speaks well, always has the answers to the teacher's questions, and spends a lot or a little time studying (sometimes people view those who study a lot as intelligent people, though intelligent people are often the ones who don't need to study).

We all have ideas about what constitutes intelligence, but it is a far more amorphous concept than it is treated in common parlance. At times, intelligence may appear to be one of those "I know it when I see it" characteristics, but efforts are still being made to describe and quantify what it means to be smart.

Street Smarts

When you were in school, you probably knew kids who were good at math but who weren't strong writers, and vice versa. Some kids excelled in music, but didn't do well in history. And some classmates probably did terrible in all their subjects, but they knew how to rebuild a car engine or how to survive if they were lost in the woods, which were things their brainy classmates would have difficulty doing.

Intuitively, we are aware of different types of intelligence. Some people are good with numbers, others with music, and still others are good with mechanical tasks. Occasionally, one person is good at several activities, but more commonly even the people we consider very smart excel in a limited number of areas.

In addition to what we think of as school smarts, which is the ability to do well in classroom subjects, we also recognize people who have what we typically call "street smarts." These people may also do well in school, but often they do not. Their intelligence relates to how they manage in life outside the classroom. They know how to get the best deals in stores, how to get from place to place on public transportation, and what to do in a threatening situation. This type of intelligence is not typically measured in any scientific way.

Athletes often are stereotyped as "dumb jocks," but most of the greats are renowned for their smarts in their sports, such as Magic Johnson or Larry Bird in professional basketball or most NFL quarterbacks. In addition, a handful of athletes are brilliant off the playing field as well, such as Rhodes scholar and former Senator Bill Bradley, who was a Hall of Fame forward for the New York Knicks.

Psychologist Howard Gardner of Harvard University popularized the notion of multiple intelligences. He pinpointed eight different types:

- **Linguistic intelligence ("word smart"):** This term refers to people who respond best to storytelling, poetry, metaphors, and other written and spoken language. An example would be a poet such as Robert Frost.

- **Logical-mathematical intelligence ("number/reasoning smart"):** People with this kind of intelligence learn best when given logical and reasoned arguments. Albert Einstein immediately comes to mind in this category.

- **Spatial intelligence ("picture smart"):** The ability to visualize and learn from images distinguishes people with this type of intelligence. The arts, such as drawing, painting, and sculpture, are their most effective learning tools. An artist such as Rembrandt might have had high spatial intelligence.

- **Bodily-kinesthetic intelligence ("body smart"):** People who like and excel at physical activity, such as dance, drama, and sports are said to have this type of intelligence. Michael Jordan is definitely body smart. Drama, mime, dance, facial expressions, role play, and physical exercise are all means of teaching the body to be smarter.

- **Musical intelligence ("music smart"):** Sounds, particularly music, aid musical learners. The composer George Gershwin undoubtedly fell in this category.

- **Interpersonal intelligence ("people smart"):** If you get along well with others, empathize with others, and enjoy working in groups, you may have this form of intelligence and probably benefit from learning in a cooperative setting. Freud is someone who at least thought he understood others.

- **Intrapersonal intelligence ("self smart"):** People who prefer to learn by themselves, engage in self-reflection, and tend toward the spiritual may be intrapersonal learners. The Dalai Lama is thought to have this kind of intelligence.

- **Naturalist intelligence ("nature smart"):** A more recently identified form of intelligence recognizes people who see patterns in the natural world. Those of us who can recognize plants and animals and live off the land fall in this category. These skills may be taught through identification and classification of not only the natural world, but the more common items in everyday life, from baseball cards to automobiles. Charles Darwin's observations about the diversity of species are an example of naturalist intelligence.

Some critics argue that Gardner's divisions are more related to personality traits and motor skills than mental capabilities, that these forms of intelligence are impossible to measure, and that research shows that people have general rather than specific intelligence. Gardner's supporters maintain that recognition of these multiple intelligences can improve the learning opportunities for students by encouraging teachers to use music, art projects, role play, field trips, and other techniques to reach students who respond better to these approaches than conventional ones.

Another question raised by Gardner's theory is whether corresponding multiple stupidities exist. We'll leave that question to others!

Incidentally, Gardner is not the only one to suggest different forms of intelligence, and the alternatives are not limited to the handful he identified. For example, a group of University of Colorado researchers suggest some people have what they call visuospatial intelligence. This refers to the ability to picture in your mind how something might look or be arranged in advance of performing a task. For example, before drawing a picture, you might have a mental image that you then commit to paper. To give a sense of how two people might have different levels of visuospatial intelligence, consider the routine task of loading luggage into a trunk. Some people can only do it after a period of trial and error, pulling suitcases in and out, rearranging them, and sometimes they simply give up and tie the trunk door down with rope. Individuals with greater visuospatial intelligence can arrange the luggage on the first try or with little effort. The researchers found that people who were good at visuospatial tasks also excelled at managing goals, performing several tasks at once, and behaving less impulsively. Interior designers and space planners have these skills.

Do Sex and Race Matter?

Gender does not seem to affect intelligence. Men and women do nearly equally well on most types of tests. Men sometimes perform better in certain areas, such as math, science, and social studies, and women excel in reading, writing, and a few other tasks. However, the overall conclusion of most researchers is that such differences in test results are based less on biology than environmental factors.

An even more controversial hypothesis is that genetics accounts for the difference in measured IQs of racial and ethnic groups. Some researchers have presented evidence suggesting a hereditary basis for the generally lower IQ results of African Americans (who score, on average, 15 points lower than whites) and, to a lesser extent, Hispanics who score about midway between African Americans and whites.

One problem with the suggestion of a genetic link between race and intelligence is the classification of race and ethnicity. Who is black? Is it someone from Africa, someone with African ancestry, someone with 50 percent African ancestry? A person can be classified as "black" in the United States if that person has any amount of black African ancestry.

The evidence for a genetic explanation for the differences in IQ scores among blacks and whites is inconclusive at best. A more common view is that the differences in IQ scores among racial and ethnic groups are based on environmental factors, such as the socioeconomic status of the family, nutrition, and the quality of education available to them.

How Smart Are You?

The discussion thus far has focused on subjective measures of intelligence. Any such judgments involve problems of perception and bias. Scientists, therefore, have attempted to develop objective tests and standards to measure intelligence. The best known is the IQ (intelligence quotient) test. Chances are you took one of these tests when you were in elementary school.

The IQ test grew out of an effort by the French psychologist Alfred Binet at the beginning of the twentieth century to develop a way of identifying students who would have difficulty learning. He devised an exam that tested memory, reasoning, problem solving, and vocabulary. Stanford psychologist Lewis Terman adapted Binet's test and changed the scoring system so that it was based on a computation of a child's mental age relative to their chronological age. For example, if a six-year-old's test results were consistent with the mental age of an eight-year-old, the IQ would be calculated this way: mental age (8) divided by chronological age (6) and then multiplied by 100 (so the number would not be a fraction) to equal 133.

The Stanford-Binet test, as it is now called, does not use this method of calculation anymore. Instead, it uses a different statistical method based on how a person's score deviates from the average of other test takers of the same age, but it retains a similar scale with 100 as average. As the following table indicates, genius level is 140. How many geniuses are there? About 1 in 125 people achieve this score on IQ tests.

Terman's IQ Test Classifications

IQ Range	Classification
140 and over	Genius or near genius
120–140	Very superior intelligence
110–120	Superior intelligence
90–110	Normal or average intelligence
80–90	Dullness
70–80	Borderline deficiency
Below 70	Definite feeblemindedness

IQ tests have fallen into disrepute in recent years for a number of reasons. One is that many people feel that such tests do not accurately examine the different types of intelligence that exist. The tests tend to be more focused on math, reading, writing, and comprehension skills and don't really measure many of the other skills associated with intelligence, such as artistic ability. In addition, some people are just better test takers than others; such people may have practiced specifically for the test or may be less affected by the stress of an exam.

A second problem is that low IQ has a certain stigma attached to it, and IQ tests were often used to sort and label people. In the past, especially, children were often treated differently depending on their IQs. Those who had low IQs might be taught at a lower level than those with higher IQs. By being treated as "dumber," the low IQ students might not be challenged to fulfill their potential, they might be branded as slow or stupid, and they might be directed toward particular fields of study and employment that weren't thought to require as much intelligence.

Certain minorities, particularly African Americans and Hispanics, have frequently voiced concern about IQ tests because they often do much worse on the tests than other groups. These differences may well be attributed to difficulties with the language, culturally or racially biased questions, and inequalities of life experience. Rather than a bias in the questions, however, it may be that the performance of these groups is primarily a result of educational and socioeconomic disadvantages.

Despite the reservations, proponents of IQ tests argue that much can be learned from the tests. They argue that people who do well on one kind of test typically do well on others and those who do poorly on one have similar difficulties on others. They also note that the tests are good predictors of student achievement in school and capability outside the classroom. Consequently, the Stanford-Binet and other newer tests continue to be used in schools throughout the United States.

What do the IQ test scores mean?

Gray Matter

The person with the highest IQ today (228) is reputed to be Marilyn vos Savant, who is in the Guinness Hall of Fame. Of course, such a high score is no longer achievable because the scoring on the tests has changed.

Individuals in the top 5 percent of the adult IQ distribution (above IQ 125) can essentially train themselves, and few occupations are beyond their reach mentally. People with average IQs (between 90 and 110) are not competitive for most professional and executive-level work, but they are easily trained for the bulk of jobs in the American economy. Adults in the bottom 5 percent of the IQ distribution (below 75) are very difficult to train and are not competitive for any occupation on the basis of ability.

Are You a Genius?

American Mensa is an organization for anyone who scores in the top 2 percent of an accepted standardized intelligence test. The following questions are an example of those asked on tests to join Mensa. The answers are at the end of the chapter. (No fair peeking!)

1. What two words, formed from different arrangements of the same seven letters, complete the following sentence?

   The foundation collapsed because the owners _____ the obvious signs that the supports were _____.

2. Start with the number of the original American colonies, multiply by the (alleged) number of feet on a centipede, and add the number of pence in "sing a song of ...". What's the answer?

3. Can you find the two bird names that are hidden in the following sentence? (The letters are in consecutive order.) Zagreb, Eve thought, is now less than a perfect vacation spot.

4. Can you think of a word that changes both number and gender when you add an *s*?

5. Watching the finish of the race, Bob observed that there were two runners in front of a runner, two runners behind a runner, and a runner between two runners. What's the minimum number of runners in the race?

6. How many common English words can you make from the following letters, using all the letters only once in each word?

   A A C G H I L L P R Y

7. If Mary has only one grandson, what relationship to her is her only son's daughter-in-law's husband?

8. What would you call metal containers full of popcorn, peanuts, or candy? _ _ _ _ k _ _ _ _

9. Can you find the names of three people that are hidden in the following sentence? (The letters are in consecutive order.) I'm angry and upset about our spat.

10. Bill flies 3,000 miles to Paris at 600 miles per hour. Because of head winds, the return flight takes eight hours. What is his speed for the entire two-way trip?

Questions copyright Dr. Abbie F. Salny. For more information about American Mensa, visit www.us.mensa.org.

And if you want to try your hand (and your brain) on another IQ test, take a look at Appendix C!

What Makes You Smart?

We humans are smarter than other members of the animal kingdom because of the size of our cerebrums. In addition, the incredible specialization of the components of our brains allows us to devote relatively more of our brain matter to thought. We have other parts of the brain that can handle other bodily functions such as balance and movement.

As to the differences in intelligence levels among us humans, some research has suggested that the speed of nerve conduction is different in smarter people and that they use less energy solving problems. Other scientists have found differences in the brain wave patters of people with different levels of intelligence.

Intelligence is to a large extent inherited. However, although smart people tend to have bright children, heredity doesn't tell the whole story. The environment can also make a difference. Children who spend a lot of time reading, writing, studying music, and pursuing other such activities can become very smart in the sense of possessing a lot of knowledge. Opportunity is also key. For example, the film *Stand and Deliver* told the story of Jaime Escalante, a math teacher at predominantly Hispanic Garfield High School in East Los Angeles, who taught his students calculus. Despite the widespread belief that they couldn't learn such a difficult subject, Escalante's students began to take and pass the advanced placement calculus test used to earn college credit in numbers that ranked Garfield among the best schools in the country.

The ability to process information, and the speed at which this information is processed, however, is largely an innate characteristic, which means you are born with it. Put another way, just as some people are born with greater potential to run fast, jump high, appear physically attractive, and achieve a certain physical height, some elements of intelligence are genetically programmed.

The way that psychologists typically try to separate the effects of heredity and environment on intelligence is by studying twins, especially those who are adopted. Because they have the same genetic makeup, the intelligence and behavior of identical twins should be similar unless it is influenced by their surroundings. In the case of intelligence, studies have suggested that differences in IQ are more closely related to genetics. For example, twins raised in different adopted homes still had similar IQs. Furthermore, researchers found that as twins became older and further removed from the direct influence of parents and teachers, the differences in their IQs become more closely related to their genetic differences. Put another way, adopted children may initially have IQs resembling their parents, but they later lose this resemblance and take on the IQ of their biological parents.

What's Your Gift?

Lots of children work hard and are smart, but those labeled "gifted" are among the top 2 to 5 percent in intelligence. They tend to work things out on their own often with original solutions, they learn more quickly and master subjects faster than other students, and they are driven to achieve.

On the negative side, gifted children are often isolated because they are labeled as nerds and often behave differently in ways that reinforce the stereotype of the eccentric genius. Once identified as gifted, children face great pressure to succeed; they are expected to eventually record extraordinary accomplishments, and they are often pushed by their parents and teachers to fulfill this potential at the expense of leading a more conventional child-hood. Gifted children often have difficulties in social situations because their peers do not have the same interests or operate on the same intellectual level. Young women who are gifted tend to have a particularly difficult time because they are generally not encouraged to speak up in class and are often discouraged from pursuing "masculine" subjects such as math and science.

The archetype of the genius is the "Renaissance man," an accolade attached to Leonardo da Vinci, who had an extraordinary range of interests and made a historic mark in both the arts and sciences. Although genius is a term that usually suggests overall brilliance, the truth is that very few people are especially capable in multiple areas. More commonly geniuses are people with special skills in a particular area, such as mathematical or musical geniuses. This unevenness may in some instance be due less to their deficiency in those areas than in their lack of interest in applying themselves to other fields. For example, a brilliant mathematician might be capable of being a great artist if he or she had the same motivation to do both.

Gray Matter

English psychologist Charles Spearman (1863–1945) found that people who did well on one type of test typically did well on others as well. He concluded that individuals have a general intellectual ability that he labeled the "g" factor. He believed we are born with this trait, which is unchanging, and determines our potential, but that people could have strengths and weaknesses in specific areas that he called "s" factors for specific kinds of intelligence.

These unevenly gifted people often do not measure as geniuses on IQ tests because their special talents are in a particular area rather than across the board, and this unevenness is penalized in test scoring. The same problem is evident in college entrance exams such as the SAT where a brilliant mathematician may do poorly on the verbal part of the test or an extraordinary artist may do poorly on both the math and verbal portions of the exam. The difficulty is often finding ways to keep these unevenly gifted people interested in the areas in which they excel while also working on their deficiencies.

For those of you concerned about whether little Johnny or Jane will grow up to be a genius, keep in mind that few people are *prodigies* such as Mozart who display their brilliance as

Words of Wisdom

Prodigies are people with exceptional talents that are usually displayed at a young age.

toddlers. Remember from Chapter 9 that the rate of development is different for everyone. Consider that studies have found that many world-class mathematicians and inventors did not learn to read before attending school and had difficulty learning to read and write.

Low IQ

Parents, teachers, and physicians sometimes discover that children below the age of 18 have difficulty adjusting socially and learning. If their IQ is below 70-75, and they have significant limitations in two or more skills required for daily life (e.g., communication, reading, writing, and social skills), they are typically referred to as mentally retarded. Approximately 3 percent of the population has mental retardation.

The overwhelming majority of people who are retarded have relatively mild conditions that make them learn new skills and information more slowly. Many can live independent lives as adults. A little more than 1 in 10 people have more severe mental retardation. Their IQs are below 50 and they require far greater assistance, but still can live productive lives without requiring institutionalization.

The cause of about one-third of the cases of mental retardation is unknown, but the remaining two-thirds are known to result from a variety of factors that affect a baby in the womb and in early childhood, including malnutrition, environmental hazards, certain illnesses during pregnancy such as syphilis or rubella, early childhood diseases such as measles or whooping cough, premature birth, smoking, and genetic disorders. Retardation can also be caused by an injury to the brain. None of these automatically result in retardation, but they can be a cause. The three most common causes are Down's syndrome, fetal alcohol syndrome, and fragile X (a gene disorder of the X chromosome that is the most frequent inherited cause of retardation).

Gray Matter

The term "mental retardation" is often considered pejorative, but it is how the condition is described by The Arc, a prominent organization that advocates on behalf of people with developmental disabilities

There is no cure for mental retardation or sure way to prevent it. Still, a large number of cases are prevented today because of more advanced screening procedures, vaccines for diseases that can cause the problem, removal of environmental toxins, improved prenatal care and diet, and the more common use of child safety seats and bicycle helmets.

The Real Rain Man

You have probably seen the film *Rain Man* or a segment on *60 Minutes* or some other television show about people with IQs that indicate a severe mental disability who can perform amazing mathematical calculations in their head or listen to a song once and then play it from memory on the piano. These people are called *savants*.

> **Words of Wisdom**
>
> The condition was first named idiot savant in 1887 by Dr. J. Langdon Down, the man for whom Down's syndrome is named. He used the word *idiot* because that was the accepted classification for the mentally retarded (IQ below 25) at the time. A **savant** (from the French *savoir*, meaning "to know") is a learned person. The word *idiot* is no longer used to describe this condition because it is considered derogatory.

Dustin Hoffman, who later played Raymond Babbitt who was the savant in *Rain Man*, was influenced by a profile he saw of Leslie Lemke on *60 Minutes*. Lemke was born prematurely with brain damage and severe eye problems that required the removal of his eyes when he was an infant. As he grew, Lemke's adopted parents noticed that he had a remarkable ability to repeat music that he heard. One night, when he was 14, Lemke heard Tchaikovsky's Piano Concerto No. 1 for the first time. He sat down at the family piano and played the piece flawlessly. Despite being blind and suffering from cerebral palsy, Lemke has a unique gift that allows him to recall and play any piece of music after hearing it just once.

Though the film character was a composite of several people, the real "Rain Man" is Kim Peek, a man born with damage to his cerebellum, no corpus callosum, and suffering from behavioral autism. Peek has read and remembers the content of more than 7,000 books, can calculate a person's day of birth and the year and the date the person will be eligible to retire, and remembers trivia on a wealth of subjects including history, movies, music, and sports.

People with such extraordinary skills in a particular area when they are otherwise autistic or otherwise mentally challenged are extremely rare, fewer than 25 such prodigies are living today. How they can do the things they can is, not surprisingly, a great mystery.

Size Matters

Do geniuses have bigger brains than the rest of us? The answer is no, but this answer does not mean that the size of the brain has nothing to do with intelligence.

IQ Points

According to Guinness, the world record for the heaviest brain was 5 pounds, 1.1 ounces. The lightest normal brain was 1 pound, 8 ounces.

Many animals are very intelligent, though they can't compare to us. Dolphins and chimpanzees, for example, have demonstrated sophisticated problem-solving and communication skills. Their brains are similar in size to humans, but as the following table shows, other animals that have brains of equal size do not have the same level of intelligence. The animals with the biggest brains are whales and elephants, but no one has shown, as yet, that either species possesses great intelligence.

You might think then that the more important issue is the relative size of the brain. For example, a human's brain is a little more than 2 percent of body weight whereas an elephant's large brain is still only .2 percent of its body weight. The only problem with this theory is that an animal like a shrew, whose brain is more than 3 percent of its body weight, should be a genius.

Average Brain Weights (in grams)	
Sperm whale	7,800
Elephant	6,000
Bottle-nosed dolphin	1,500–1,600
Adult human	1,300–1,400
Camel	762
Horse	532
Gorilla	465–540
Cow	425–458
Chimpanzee	420
Newborn human	350–400

The brain also shrinks as we age. By age 70, the brain loses 5 percent of its mass, and the figure can grow to 20 percent by age 90. Intelligence, as measured by an IQ test, decreases in the elderly individual; however, this decrease may be related to multiple factors in addition to brain cell loss, such as the speed with which the test can be finished and problems related to memory.

What about differences in the brains of men and women? On average, women's brains are slightly smaller, but that is also in proportion to the rest of their bodies, which are also typically smaller than those of men.

Two Brains Are Better ...

One way that researchers have investigated the role of the brain in intelligence has been to examine split-brain patients, that is, people whose corpus callosum has been cut so that the two hemispheres of the brain are no longer connected. These people looked and acted like anyone else, but Nobel Prize-winner Roger Sperry conducted a number of experiments that revealed fascinating differences in the way the two halves of the brain work and the impact the split had on the patients.

In one experiment, a patient was told to stare at a screen. A picture of an object, say a fork, was flashed on the right side of the screen. The light waves entered the eye and crossed the optic chiasm into the left hemisphere where the language center was located. When asked what he or she just saw, the patient readily responded, "fork." If the picture of the fork was flashed to the left of the screen, however, the signal traveled to the right hemisphere. When asked again what was flashed on the screen, the patient said that he or she didn't see anything because the right side of the brain couldn't "talk."

In another experiment, patients were shown a picture referred to as a chimera (a picture with parts that don't match). The left half was the face of a woman, and the right was that of a man. When patients were told to focus on a spot in the middle of the figure, the visual information about the woman's face traveled to the right hemisphere, and information about the man's face went to the left hemisphere. If the patients were asked whether the picture was a man or a woman, they would say it was a man (the image of the male side went to the language center on the left). However, if the patients were told to choose a photograph of the face that was flashed on the screen, they would usually choose a woman's picture, suggesting that the right hemisphere was dominant for recognition of faces. In another example, a blindfolded split-brain patient who was given an object in his or her left hand could identify it verbally, but if the same object were put in his or her right hand, the person would not be able to identify it.

The result of the split-brain studies was the conclusion that the two hemispheres controlled different elements of thought and behavior. The left brain was crucial to language and speech, and the right side was dominant for visual-motor tasks. When the corpus callosum is intact, the two hemispheres work together.

Da Vinci, Mozart, and You

Creativity is the capacity to generate new ways of solving problems and to be original and imaginative. Where does this trait come from that allows a Michelangelo to produce the painting on the ceiling of the Sistine Chapel or an Edison to invent the light bulb or a Michael Crichton to imagine a Jurassic Park or a George Lucas to create a galaxy far, far away?

Nobody knows.

Interestingly enough, the question of what makes people creative is one that people thought they did know the answer to for most of history. The answer was plain as day: Creativity comes from God. After all, the word *inspiration* literally means "to be breathed on by gods."

Even today, many people would probably say that creativity has some divine origin, but scientists are still seeking a more concrete explanation that would identify characteristics of the brain that might allow certain people to think "outside the box." Could it be, for example, that creative individuals have different circuitry in their cortexes that allow them to combine information in more ways than the average person? Some researchers believe that people have different filtering systems so that "crazy" ideas that might be discarded by the brains of others are retained in the minds of geniuses. We just don't know how it all works yet. Maybe we never will.

Psychologists offer their own ideas about creativity, but these explanations do not describe the physical process involved. For example, Freud thought creativity was a product of unconscious drives, particularly sexual drives.

Creativity is also seen as a process involving at least four stages: preparation, incubation, illumination, and verification. First, creative people see a problem or goal and think about whether it is something they want to pursue and whether they have the knowledge, materials, and other assets needed to move forward. Next they consider the different possibilities. Artistic people tend to be free and expressive in their internal deliberations while scientists are more disciplined and logical in their analyses. The third phase is what some refer to as the "Eureka! Moment" when creative people see the answer or the approach they should take. Finally, the verification stage involves the tweaking necessary to put the discovery or idea into its final form.

Einstein's Brain

If someone asked you who the smartest person who ever lived was, chances are the first name to come to mind would be Einstein.

Being a genius, Einstein recognized that there might be something special about his brain and instructed others to save and study his brain after his death (the rest of his body could be cremated). This gift was quite remarkable when you think about it. After all, scientists can't normally study the anatomy of the brains of living geniuses because they are being used, and most geniuses do not donate their organs to research after death.

Einstein's generosity proved not to be as straightforward as you'd expect. After he died on April 18, 1955, at the age of 76, pathologist Thomas Harvey removed Einstein's brain at Princeton Hospital. Then, mysteriously, Einstein's brain disappeared.

In 1978, Steven Levy, a reporter for *New Jersey Monthly*, was assigned to find Einstein's brain. He tracked down Dr. Harvey, who had moved to Wichita, Kansas. Reluctantly, Harvey admitted that he still had Einstein's brain, and it was sitting in his office. Harvey opened a box labeled Costa Cider and pulled out two mason jars that contained the most esteemed brain the world had ever known. Harvey had sectioned (sliced) most of the brain, leaving the cerebellum and part of the cortex intact.

Harvey did not find anything unusual about the structure of Einstein's brain, but after Levy's discovery was publicized (and it caused quite a sensation!), other researchers were allowed to study the great scientist's brain. One paper found that Einstein had a greater number of glial cells per neuron in one particular area of his brain. The authors concluded that these neurons required and used more energy and, therefore, might have improved Einstein's thinking and conceptual skills. Another study found that Einstein's brain was smaller than average in weight but had a greater density of neurons in the cortex. Yet another study found that Einstein's brain had an unusual pattern of grooves in the area of the parietal lobes thought to be involved in "higher" mental functions including writing, spelling, and calculation. His brain was also wider than other brains that it was compared to. The conclusion of that paper was that Einstein's brain might have had better connections between neurons used in math and spatial reasoning.

None of these studies have been conclusive, and each has methodological problems. For example, the studies compare the greatest mind in history with "normal" brains from people of different ages. To see whether the features of Einstein's brain are unique to math geniuses, for example, you would need to compare them to those from the brains of other math geniuses. To a large extent, all these comparisons involve apples and oranges.

IQ Points

Einstein is not the only famous person to have his brain preserved and studied. After Vladimir Ilich Lenin, the man who led the Russian revolution in 1917, died in 1924, the Soviet government hired German neuroscientist Oskar Vogt to study Lenin's brain. Vogt found that the revolutionary's brain had unusually large and numerous neurons in one area of his cerebral cortex. Vogt had to focus on Lenin's right hemisphere because the left was too seriously damaged by the cerebrovascular disease that had killed him. Vogt believed these cells were related to "associative thinking," and that this structural peculiarity might account for the strikingly acute and penetrating mental processes that had characterized Lenin's personality.

The 10 Percent Myth

Albert Brooks made an hysterical movie called *Defending Your Life*. The premise of the film is that after death you go to a kind of way station where your life is evaluated before it is determined where you will go next. Actor Rip Torn plays the person defending Brooks's life and makes frequent reference to the "fact" that humans use only a tiny percentage of their brains (he says 2 to 5 percent), but that he and the others who have evolved beyond the earthling stage use roughly half their brains and are therefore intellectually superior to Brooks.

This part of the film reflects a common misconception that humans only use a fraction (typically the figure used is 10 percent) of their brains. No one knows exactly where this myth came from, but there is certainly a widespread notion that we don't use our full mental capacity, but that if we did, like the Torn character in *Defending Your Life*, we could have superhuman mental powers.

Fugettaboutit!

If it were true that we didn't use all of our brains, those unused areas would likely deteriorate, and the number of neurons would decrease. Also, from the related evolutionary standpoint, why would human brains grow as humans evolved if we weren't going to use them to their fullest? Evolution would dictate that the brain would then become smaller, not larger.

Brain imaging research has also shown that most of the brain is working at any given time. As we've learned in prior chapters, virtually every function is spread across multiple areas of the brain, so reading, writing, or speaking activates several regions of the brain. The parts of the brain that are not used in those particular activities will inevitably be involved in other mental and physical functions during the course of any given day.

In addition, we know that damage to any part of the brain causes some impairment. Sometimes even a tiny lesion, if it interrupts important neural pathways within the brain, may produce severe deficits.

The truth is that we use 100 percent of our brains. Recent studies with PET scanning have shown that different parts of the brain are working harder during certain circumstances, such as solving a mathematical problem, but other parts remain active simultaneously in controlling vital functions.

Mad Genius

Why is *genius* so commonly associated with mental illness? The answer is that we know of many famous people who were considered geniuses in their fields who went mad or behaved in bizarre ways or suffered from some debilitating mental problem. Consider the biographies of people such as van Gogh, Tolstoy, Tchaikovsky, and Poe.

Words of Wisdom

The word **genius** comes from the ancient belief that men of genius had a guardian spirit or demon (*genio*) that whispered in their ear.

One condition that has often been associated with creativity is manic depressive disorder (described under bi-polar disorder in Chapter 20). While in the manic mode, people have tremendous energy and can work long hours. They also tend to be less conscious of conventions, social and otherwise, so they can think "outside the box." When the manic phase passes, however, they can plunge into deep depressions.

Scientists have conducted a number of studies to determine whether a relationship exists between creativity and mental illness. The results show that creative people do seem to have a higher incidence of psychological problems, but scientists have not proved that one causes the other.

The influence of drugs on creativity is also controversial. Some artists are convinced that they need drugs such as marijuana to free their minds to produce original thoughts. On the other end of the spectrum, creative people with mental illness, particularly bipolar disorder, often believe taking medication to treat their illness has an adverse impact on their ability to think.

In the book and film *A Beautiful Mind* about the life of Nobel laureate John Forbes Nash Jr., the mathematician is portrayed as a genius who had the ability to see patterns in numbers that no one else could see. He also was delusional and suffered from schizophrenia. He was given medication to control his illness, but Nash was convinced that the drugs made it impossible for him to continue his work. He ultimately stopped taking the medication and returned to creative work, but he was also constantly battling to keep his behavior in check. Later, newer medications allowed him to work and control his illness. As critic Roger Ebert wrote in his review of the film, Nash's story left us with a question: "Did his ability to penetrate the most difficult reaches of mathematical thought somehow come with a price attached?"

Formula for Thought

Parents are always looking for an edge for their children and, after all, we all want the best for them and hope they'll be not only good people, but geniuses to boot. Well, one of the newest ideas for building baby brain power is to feed them formula that is supplemented with two fatty acids (docosahexaenoic or DHA and arachidonic or ARA) that are found in breast milk and are believed to be associated with brain and vision development.

These premium formulas have just come on the market and are, naturally, more expensive than the garden variety on the shelf, and some experts still question whether they will have any beneficial impact. The Food and Drug Administration did certify the supplemented formulas are safe, which allowed the manufacturers to sell them, but the FDA has not concluded they are better for babies. So far, there appear to be no harmful effects from the supplements, so many parents feel there's not much to lose and want to give their babies whatever most closely approximates the nutritional content of breast milk.

Smart Pills

The desire to improve memory and get smarter leads people to try all sorts of strategies, including taking different types of drugs and herbs reputed to have beneficial effects on memory. Some of these affect particular structures in the brain or neurotransmitters. Some medications, and various herbs and vitamins, have been shown to improve memory. You should discuss these supplements with your physician before taking them, because they may not be compatible with medications that have been prescribed for you.

And even if you could pop a smart pill, would it be a good idea? For example, would such a drug change your personality? Should restrictions be placed on who gets the pills? Should you be allowed to pop one before a test?

These questions get into difficult issues of medical ethics and philosophy, which are beyond the scope of this book. The point is that, like the debate over human cloning, medical advances that affect the mind may not be unambiguously beneficial.

Here are the answers to the Mensa quiz:

1. Ignored, eroding.
2. 1306 ($13 \times 100 = 1300 + 6$).
3. Grebe, owl.
4. Princes to princess. (You may find others.)
5. 3
6. Calligraphy and graphically are the words we found. Good for you if you found another.
7. Her grandson.
8. Snack cans.
9. Ima, Ryan, and Pat. (Tab is not really a name.)
10. Approximately 461 miles per hour. (Five hours to Paris plus eight hours to return equals 13 hours for 6,000 miles. Six thousand divided by 13 equals 461.5.)

The Least You Need to Know

◆ People can have all different types of intelligence; few people are good at everything, and even many exceptionally bright people are unevenly gifted.

◆ Though test results often show differences in the intelligence of men and women and among ethnic and racial groups, most evidence does not support the idea that intelligence is based on gender or race.

◆ IQ tests are imperfect and have sometimes been misused, but they are still good predictors of student achievement.

◆ The idea that humans use only a fraction of their brains is a myth. You use all of your brain.

Part 5

The Sick Brain

Brain injuries, diseases, and mental disorders are serious, often debilitating, and sometimes deadly problems faced by millions of people, and we explore them in this section with as much detail and sensitivity as they deserve. Because so many brain diseases and injuries exist, we've tried to focus on the more common and well-known problems in the broad categories of brain injuries, degenerative disease, neurological disorders, alcohol and drug abuse, and mental illness. We offer detailed descriptions of these conditions, but these descriptions are not meant to be used for diagnosis, especially self-diagnosis. Nothing in this book is meant to substitute for the judgment of a physician.

Chapter 17

The Computer Crashes

In This Chapter

- Bumps and breaks in the head
- The truth about tumors
- Brain infections
- Passing on disease
- Stroke is no joke

The brain is a very delicate organ, and despite the protection of the skull, it can be damaged in a variety of ways. The brain, like the rest of the body, is also susceptible to a number of diseases. This chapter describes some of the most common injuries and diseases that affect the brain. Keep in mind that we're only going to cover the most common; many more possibilities exist.

In addition, these descriptions of often complex conditions are brief, and you should not use them for diagnostic purposes. If you believe that you suffer from any of the symptoms described in this chapter, don't panic! There are hundreds of other possible causes for the symptoms. Consult your doctor for an assessment.

Trauma: More Than a Little Bump

Trauma to the brain occurs in many ways and causes a variety of problems. Injury may be very minimal and have no significant consequences or may be major causing one or more severe neurological deficits. In extreme cases, an injury can cause a person to slip into a prolonged coma or vegetative state. The individual in the latter condition is aware of his or her surroundings but cannot communicate.

When the Brain Causes Pain

Injury or damage to one of the peripheral nerves may cause paralysis of the muscles to which the nerve goes or neuropathic pain, which is a strange burning or shooting pain,

> **Words of Wisdom**
>
> **Shingles** is a skin rash caused by the same virus that causes chicken pox. You get shingles because of the reactivation of the herpes varicella (zoster) virus.

that may become chronic. One particularly unfortunate example of this condition is known as "phantom pain," which can occur in patients who have lost a limb or part of a limb. In those cases, patients experience severe pain in a surgically or traumatically amputated limb or distal part of a limb such as a hand or foot. The sensation may be so intense that the individual is convinced that the amputated part is still present. Diabetes or *shingles* may cause another kind of neuropathic pain. Neuropathic pain is extremely difficult to treat.

Butting Heads

Head injury that damages the brain is probably better characterized as brain injury. A brain injury may be the direct result of trauma, such as being hit in the head, or may be a secondary result, such as what occurs when too little oxygen reaches the brain for any reason (drowning, cardiac arrest, stroke, and so on).

According to the National Institutes of Health, someone receives a head injury in the United States every 15 seconds. Of this group, one dies and another becomes disabled every 5 seconds. More than two million individuals suffer a traumatic brain injury every year, and brain injury is the leading cause of death and disability in children and young adults. Head injury represents 10 to 15 percent of all injuries nationwide. From 500,000 to 750,000 head injury patients require hospitalization, and 75,000 to 100,000 die as a direct result of the head injury or related complications.

> **Gray Matter**
>
> Males are twice as likely as females to suffer a head injury. This statistic is probably related to the fact that more males ride motorcycles and bicycles than females.

Motor vehicle accidents account for 50 to 60 percent of head injuries; falls account for the next largest amount followed by assaults and gunshot wounds. Of the head injuries related to transportation, motorcycle accidents account for 20 percent, and bicycles account for about 10 percent. Alcohol plays a role in about 60 percent of head injuries. The types of head injuries vary from minor (80 percent) to moderate (10 percent) to severe (10 percent).

One of the most common head traumas is called *concussion*. Concussion is characterized by a temporary loss of consciousness for a period of a few seconds to a few minutes without structural pathologic changes in the brain. A closed head injury means there is no scalp laceration or underlying skull fracture. Athletes engaging in contact sports often have concussions. Because the effects of concussions are cumulative, athletes who have had multiple concussions are advised to give up the sport because of the danger of permanent brain damage. Professional football quarterbacks Steve Young of the San Francisco '49ers and Troy Aikman of the Dallas Cowboys were forced to retire after suffering repeated concussions. New York Rangers hockey star Eric Lindros has had seven diagnosed concussions, but he continues to play.

> **Code Blue**
>
> The number of head injuries and the number of fatalities related to transportation have been significantly reduced by the use of seat belts and air bags in cars and by the use of helmets by motorcycle and bicycle riders. We can't say it strongly enough: Always wear your seat belt in the car and a helmet on a bike or motorcycle!

In an article for the Connecticut State Medical Society Sports Medicine Bulletin, Dr. Carl Nissen offered some guidance for evaluating and treating concussions that occur during sporting events. He divided concussions into four types: hard head knocks or "Bell Ringers"; mild injuries that cause persistent, but limited headaches and loss of some physical and mental abilities; and two levels injuries that result in the loss of consciousness, one more severe than another.

Types of Concussions

Bell Ringer	Grade 1	Grade 2	Grade 3
Hard head knock, no amnesia	No loss of consciousness, minimal retrograde symptoms may last more than one minute	Loss of consciousness for less than 15 seconds, mild retrograde amnesia only	Loss of consciousness for more than 15 seconds

For most people, treatment is relatively straightforward, rest and relaxation. For athletes, however, the issue is more complex because of the desire of coaches and players to return to their sports as quickly as possible. The following table shows the general guidelines Dr. Nissen suggests, with the caveat that it is better to err on the side of caution.

Guidelines for Athletes Suffering Concussions

Bell Ringer	Grade 1	Grade 2	Grade 3
Return to play in 15 minutes if mentally clear and no symptoms with provocative tests	Return to play in 15 minutes if mentally clear and no symptoms with provocative tests	Return to play that day usually not possible, return in one week if no ongoing symptoms	Return to play after clearance from knowledgeable personnel only, minimum one week

Usually an athlete can resume activity after 15-20 minutes of rest following one of the milder injuries. First, it is important to perform a mental status test and physical challenge. If the player passes the tests and has full range of motion and strength in the upper body and neck, then it should be safe to return to the sport. It's also advisable to monitor the athlete for the next day or two.

In the case of more serious injuries resulting in a loss of consciousness (Grade 2), the athlete needs to be closely watched, especially if they've had previous concussions. No one should be allowed to play after suffering two concussions in one day. The most serious concussions (Grade 3) require urgent care and athletes should be immediately transferred to a hospital for evaluation and treatment. Usually this will include, at a minimum, x-rays of the head and neck and perhaps a CT scan. If symptoms, such as loss of coordination, inability to focus, headaches, irritability, or memory loss persist after any of the concussions—even the seemingly mild ones—further evaluation is warranted.

Code Blue

> In addition to the physical damage to the brain that can result from concussions, recent studies have shown that people who suffer concussions and other head injuries in early adulthood may be more prone to depression.

Breaking Your Head

A skull fracture often represents a more serious head injury, although in some cases the fracture of the skull may absorb most of the energy of the injury and the brain may be less traumatized than in a closed head injury without a fracture. Many of the more serious head injuries involve bleeding inside the skull that causes a blood clot, called a *hematoma*, on the surface of the brain or within the brain substance itself. If the period of unconsciousness is prolonged or occurs after a lucid interval (a period of time during which the individual is awake and alert and neurologically intact), a traumatic intracranial hemorrhage should be suspected.

A blood clot in the brain tends to be more serious than a clot forming in another part of the body. If you have a blood clot in your arm, for instance, the skin can expand so that pressure from the blood clot will eventually stop the bleeding. Because the skull is a fixed bony box, however, the tissue has no room to expand to accommodate the clot. The result is an increase in pressure inside the skull, referred to as increased intracranial pressure. If severe enough, this pressure may cause permanent neurological damage, such as blindness, paralysis, or even death.

The treatment for increased intracranial pressure caused by an expanding blood clot or a swelling brain is medication, if there is not an immediate response to the medicine, emergency surgery is usually necessary. Surgery for the removal of an intracranial hematoma is fraught with considerable risk because there may already be irreversible brain damage caused at the time of the injury. In many cases, however, the operation may be a life-saving procedure because the increased intracranial pressure is relieved. The surgery involves making an opening in the skull and then removing the blood clot (generally) with a suctioning device similar to that used by a dentist on teeth.

The final outcome from a serious head injury may be in doubt for many months and is dependent on the degree of injury and the response to medical and/or surgical treatment. Families may have guilt, denial, anger, and shock as normal reactions to the situation and often don't recognize that everything possible is being done for the patient with the head injury.

Rehab

In cases of severe head trauma, rehabilitation starts very early in the recovery process and may even begin while the patient is still in a coma (unconscious) to maintain mobility and flexibility in joints and to prevent contractures (an abnormal shortening of muscles or tendons that may permanently affect the ability to use them).

Of those individuals considered to have a minor (defined as one that does not require surgery) head injury, 70 percent will have long-term disabilities, and 60 percent will have impaired memory. As many as 10 percent will have permanent neurological deficits and may never return to work. However, with current efforts involving comprehensive rehabilitation, more of these disabled individuals are returning to the work force today than in previous years.

IQ Points

The economic burden for the country as the result of head injuries is enormous. The direct and indirect medical costs of rehabilitation, support services, and lost wages approaches an estimated 25 billion dollars annually. The average hospital stay for a patient with a head injury is 45 to 60 days at a cost of more than $350,000.

One other long-term problem related to head injury is seizures: Almost 5,000 new cases of seizure disorders related to the original head injury do not present themselves until weeks, months, or years after the original injury. Seizures of this type probably develop as scar tissue forms at the site of the brain injury. We'll have much more to say about seizures in Chapter 18.

Two Scary Words: Brain Tumor

To most people, one of the worst health nightmares imaginable is finding out that they have a brain tumor. This feeling is understandable because more than 50 percent of discovered brain tumors are malignant and carry a poor prognosis. Yet a significant number of brain tumors are benign and may be completely cured with appropriate treatment.

A brain tumor develops when brain cells do not divide properly and instead form abnormal growths. The following sections describe different kinds of brain tumors.

Skull Tumors

Tumors of the skull are usually benign and rarely require surgical removal except for cosmetic reasons. Malignant tumors of the skull rarely occur by themselves, but instead stem from another cancer, most often a malignant tumor of the breast. These skull tumors are usually blood-borne and are often associated with growths in the lung as well. Malignant skull tumors are usually treated with radiation therapy.

Meningiomas

Tumors involving the meninges (the covering over the brain) comprise about 15 percent of all brain tumors and are referred to as *meningiomas*. They are usually benign and can be completely removed surgically in most cases. The signs and symptoms of meningiomas are directly related to their anatomical location. Because these tumors grow very slowly, symptoms may be present for a long time before a diagnosis is made. For example, a meningioma growing in the region of the frontal lobe may grow to an enormous size (the size of a baseball) before the patient comes to a physician.

IQ Points

According to the Guinness Book of World Records, the largest brain tumor weighed 1 pound 5 ounces and was successfully removed from a four-year-old child.

Symptoms of a meningioma include progressively increasing headache, difficulty with vision, weakness on one side of the body, decreased memory, difficulty with speech, personality changes, and seizures. Further diagnostic evaluation is usually required because these symptoms are not exclusive to a meningioma. They may be present in cases of other types of tumors or with other types of lesions, such as brain abscesses.

Cranial Nerve Tumors

Tumors involving the cranial nerves are not common, but when they occur they present very specific clinical pictures. For example, a tumor of the optic nerve causes progressive visual loss and headache. These rare tumors are found more often in children than adults. They may be treated with radiation, but treatment may also require removal of the eye.

The most common cranial nerve tumor is that of cranial nerve VIII, which is the vestibulocochlear (auditory) nerve that deals with balance and hearing. Such tumors are referred to as "acoustic tumors," and they make up 5 to 10 percent of all intracranial tumors. An acoustic tumor is almost always benign.

Symptoms of an acoustic tumor include deafness in one ear, headache, and tinnitus (ringing in the ear). If the tumor is very large, it may also cause decreased sensation on the same side of the face.

Very large tumors may also cause problems related to adjacent cranial nerves, including V, VI, VII, and X. Because acoustic tumors grow in the cerebello-pontine angle where the cerebellum and pons come together at the base of the brain, signs related to compression of the cerebellum, such as difficulty walking, may occur as well. If the tumor continues to grow, the ensuing compression of the brain stem will ultimately cause death.

Acoustic tumors can be completely removed by a surgical team comprised of a neurosurgeon and neurootologist (a specialist in hearing loss due to neurological conditions). Unfortunately facial nerve weakness and partial paralysis of the facial muscles is a complication of the removal of large tumors (more than 4 centimeters in diameter). More recently, excellent results have been obtained using the Gamma knife, which is a radiation source used without open surgery, to treat the tumor.

A tumor of cranial nerve V (the trigeminal nerve) usually causes numbness in the face and weakness of chewing on the same side of the tumor. It is usually treated with radiation therapy or surgery.

The Baddest of the Bad: Gliomas

A *glioma* is a tumor made up of the supporting cells (the glial cells) of the neurons. Gliomas may appear anywhere in the brain and make up about 50 percent of all brain tumors. They are generally malignant, but they vary in degree of malignancy. Astrocytoma is the most common of these tumors, and like most tumors, it is graded on a scale of I to IV, with Grade IV being the most malignant. Grade IV astrocytomas, also known as glioblastoma multiforme, are the fastest growing and most aggressive of all brain tumors.

Medulloblastoma is a highly malignant tumor seen in infants and young children. It comprises 16 percent of all cerebellar tumors. The tumor arises from the cerebellum and often fills the fourth ventricle. It may invade the floor of the fourth ventricle, making its complete removal impossible (since this is the area of the medulla and the seat of control of vital functions of the body, such as respiration, blood pressure, and heart rate). The child with this tumor experiences increasing lethargy, vomiting, headache, and gait disturbance. Because the tumor may block the normal flow of cerebrospinal fluid, the child may suffer from acute hydrocephalus (enlarged ventricles and increased intracranial pressure). The child is acutely ill and presents a surgical emergency. Treatment consists of surgical removal of the tumor with relief of the increased intracranial pressure. If the hydrocephalus is not relieved by tumor removal, a shunt procedure may be necessary.

For instance, a six-year-old child was taken to his pediatrician because of a two-week history of headache, vomiting and unsteadiness leading to several falls. A neurological consultation was obtained and an MRI scan was done, which showed a large cerebellar tumor with mild ventricular dilation. Emergency surgery was performed during which a small tube was introduced through the brain into the right ventricle because the CSF pressure was slightly elevated. The surgeon began to remove the tumor through the base of the skull, but discovered it was sticking to the fourth ventricle. He succeeded in removing 90-95 percent of the tumor and the child made an uneventful recovery from surgery. Within 48 hours the CSF pressure returned to normal and the tube in the ventricle was removed. The child did well for six months but then returned because of increasing headaches over a period of three days. An MRI scan revealed a recurrent medulloblastoma compressing the brain stem and severe hydrocephalus. A shunt procedure was done to divert the spinal fluid and all the child's symptoms cleared up. Alas, the story does not end happily; four months later the child died.

Treatment of gliomas continues to challenge even the brightest of researchers. Currently, treatment uses several techniques combined: usually, surgical removal followed by radiation therapy and chemotherapy.

Gliomas are extremely difficult to remove because they usually have tentacles that reach into deep parts of the brain. Robotic surgical techniques combined with multidimensional imaging techniques have improved the surgical results, but the results are still far from perfect. Newer approaches being investigated in many research centers include immunotherapy and gene therapy.

Pituitary Tumors

Tumors found in the pituitary gland constitute about 15 percent of intracranial tumors and are usually benign. They often grow large enough to expand out of the bony compartment of the skull that contains the normal pituitary gland. When this happens, the tumor may cause compression of the optic chiasm (remember, that's where part of the optic nerve crosses to the opposite side), and this compression results in the loss of visual fields in both eyes (called a bitemporal hemianopia). The example of the driver going down the middle of a one-way street and not being able to see the parked cars on either side represents this type of condition.

The most common symptoms of a pituitary tumor are caused by the disturbance the tumor causes to the regular production of hormones, which is controlled to some degree by the pituitary. Women may have amenorrhea (no menstrual periods) and galactorrhea (a milky breast discharge), and skin changes and alterations in hair distribution, especially in the axilla (underarm) and the pubic areas, may occur. Men with pituitary tumors don't experience such specific hormonal changes. Men may have gynecomastia (enlarged breasts), decreased libido, and impotence.

A tumor of the pituitary in childhood may cause the pituitary to overproduce growth hormone, which may lead to two conditions called giantism (extreme tallness) and, later in life, acromegaly (in which the individual has large hands, a large skull, and, eventually, enlargement of the heart, which can lead to heart failure in advanced untreated cases).

Gray Matter

Prolactin-secreting pituitary tumors, which cause amenorrhea and galactorrhea, are a significant cause of infertility in women.

Doctors can diagnose a pituitary tumor with certainty by using modern imaging technology such as the MRI. Treatment of a pituitary tumor depends on a number of factors, including the age of the patient, the severity of the symptoms, and whether the vision is impaired. Medical treatment with the drug bromocriptine may help to either eliminate the tumor or to shrink it before surgery.

Another unusual condition, sometimes seen in patients with pituitary tumors, is known as Cushing's disease (named after Chapter 4's Harvey Cushing). An individual with this disorder may have a moon face, abdominal striations, increased skin pigment and hair, increased blood pressure, and decreased sexual function.

Almost all pituitary tumors are removed through a transnasal approach (through the nasal cavity) rather than with an opening into the skull above the hairline unless the tumor is massive or has invaded surrounding structures. In some cases the Gamma knife may be used to avoid an open neurosurgical procedure. Surgical removal followed by radiation therapy provides a cure in almost every case.

Catching a Brain Disease

The brain is as susceptible to infections by virus, bacteria, and other pathogens as any other part of the body, but infections occur less often in the brain because of the blood-brain barrier. An infection that causes inflammation of the covering of the brain and spinal cord is called meningitis. When it affects the substance of the brain, the result is encephalitis. The following sections describe some of the different infections that affect the brain.

Bacterial Meningitis

Acute purulent meningitis is a disease caused by a bacterial infection such as meningococcus, streptococcus, staphylococcus, or pneumococcus. (A less virulent form of meningitis may be seen with tuberculosis or fungal infections, such as histoplasmosis and, in the San Joaquin Valley of California, coccidiomycosis.) The symptoms of acute bacterial meningitis may develop rapidly and progress very quickly. The disease can be life threatening and requires emergency treatment.

The infection may gain access to the central nervous system from an infection anywhere in the body, but an open (compound) skull fracture or an infection of the paranasal sinus is more likely to be the source of the infection of the central nervous system.

The diagnosis of meningitis requires the immediate performance of a lumbar puncture (spinal tap) and evaluation of the cerebrospinal fluid that is obtained. The evaluation should include testing of the sugar level, protein level, chloride level, cell count, and microscopic observation for the presence of organisms (bacteria). In the case of purulent meningitis, the sugar level is lower than normal, the protein level is higher, and the chloride level is lower. The cell count is dramatically elevated.

The signs and symptoms of acute purulent meningitis include progressively increasing headache, nausea, vomiting, stiff neck, and back pain, accompanied by fever and chills. A skin rash may be present if the meningococcus organism is the cause of the infection. A decrease in the level of consciousness leading to stupor and progressive coma may occur in rapidly progressive cases, which could cause death. On occasion convulsive seizures may be seen.

On examination, the physician will find an individual who appears acutely ill, who may be confused, stuporous, or semi-comatose, and who may (or may not) have a fever of up to 101 to 103 degrees. The pulse and respiratory rate are increased. Blood pressure is usually normal. There is almost always a stiff neck present.

Treatment for acute bacterial meningitis must be started even before any of the laboratory results are known. Antibiotic treatment must be started immediately when there is a strong suspicion of the diagnosis and then may be modified when the causative organism is identified.

If bacterial meningitis is untreated or inadequately treated, the effects may be catastrophic. These effects may include paralysis of extraocular muscles, deafness, blindness, hemiplegia (paralysis of one side of the body), a seizure disorder, and mental deficiency.

Gray Matter

In the neonatal (immediately after birth) period, the E.coli organism is the most common cause of meningitis, and H. influenza can cause the disease in older children.

Abscessing

A collection of material that is full of pus within the substance of the brain is known as a brain abscess. A relatively rare occurrence, a brain abscess usually develops after an infection erupts somewhere else in the body. Sometimes, the contamination is direct, such as when the infection is in or near the brain, but it may also be indirect when the blood brings the infecting agent to the brain from another part of the body (often the lungs). Awareness of a mastoid infection (which usually comes from an ear infection), and the potential for brain complications, should lead the patient to an early visit to the doctor because of the possibility it can develop into a brain abscess.

Symptoms include those related to increased intracranial pressure (headache, nausea, vomiting, and decreased level of consciousness) and seizures. Body temperature is usually normal or only slightly elevated. Once a doctor suspects a brain abscess, he or she can establish the diagnosis by using an MRI scan.

Encephalitis

Encephalitis is the result of a viral infection in the brain. In general, encephalitis is not very common in the United States, so we'll just mention a few types. Eastern, Western equine, and Japanese B encephalitis all give rise to similar clinical pictures. Symptoms may come on suddenly and include progressive stupor, seizures, headache, vomiting, stiff neck, and high fever. Cranial nerve palsies and other neurological signs are common symptoms as well.

Treatment for encephalitis depends on the organism causing it. If it is bacterial, the doctor will probably prescribe an antibiotic. If it is viral, as it is in most instances, you only have to treat the symptoms because there is no good treatment for a viral encephalitis.

Gray Matter

Effective research has totally eradicated polio, probably the most well known of the viral infections of the central nervous system, from the industrialized world through the use of the Salk and Sabin vaccines. This gives us hope that other viral diseases, such as AIDS, may be brought under control and eliminated through research efforts.

Rabies

Rabies is an important virus because sporadic cases in humans still occur when infected bats (or, more rarely, dogs and other animals) bite them. The incubation period (the time during which there may be no overt symptoms) may be many months. The clinical picture of rabies is extremely rare because most, if not all, individuals exposed to the disease are treated with anti-rabies serum (a painful but necessary treatment).

Gray Matter _____

Since 1990 the number of reported cases of rabies in the United States has ranged from one to six cases annually. In 2000, more than 7,000 cases of rabies in animals were reported (the highest percentage was in raccoons), but no human cases were reported. Almost all human rabies cases in the United States have come from bats. Outside the United States, the disease is usually the result of the bite of a rabid dog. Incidentally, Hawaii has never had an indigenous case of human or animal rabies.

Symptoms of a full-blown case of rabies include apathy, drowsiness, headache, and anorexia, followed by a state of excitability with localized twitching and convulsions. Delirium with hallucinations may also result. Attempts to eat or drink result in a profuse flow of saliva and spasms of the pharynx and larynx. The body temperature may rise to as high as 107 degrees. Left untreated, rabies can result in paralysis, coma, and death as a result of paralyzed breathing muscles.

According to the Centers for Disease Control (CDC), the number of human deaths in the United States attributed to rabies declined in the last century from 100 or more each year to an average of one or two each year. This decline is attributed to the animal control and vaccination programs begun in the 1940s that have almost wiped out rabies in domestic dogs and the development of effective vaccines and immunoglobins (proteins that provide immunity).

Because the most serious danger of rabies to humans comes from bats, keep in mind the recommendations of the CDC, starting with this one: Never handle a bat! The CDC also recommends the following:

> You cannot tell if a bat has rabies just by looking at it … so if you are bitten by a bat—or if infectious material (such as saliva) from a bat gets into your eyes, nose, mouth, or a wound—wash the affected area thoroughly and get medical attention immediately. Whenever possible, the bat should be captured and sent to a laboratory for rabies testing.

Other Brain Infections

Rocky Mountain spotted fever can cause encephalitis as a direct result of a tick bite, which transmits the organism (Rickettsia rickettsi) responsible for the disease. This disease is not uncommon in areas where outdoor camping is a popular recreation. Wearing high boots and protective clothing decreases the chance of getting bitten by the tick. Lyme disease is also caused by ticks, but it does not affect the brain.

Syphilis caused by the organism, called a spirochete, Treponema pallidum, may involve the central nervous system. The disease may affect the substance of the brain and cause symptoms of personality change, convulsions, and mental deterioration. Other symptoms will be present if only the meninges (the covering over the brain) or the blood vessels of the brain are involved. Central nervous system involvement with syphilis was common before and after the turn of the twentieth century. However, since the end of World War II, improved methods of prevention and treatment have caused a very significant drop in the number of cases being seen.

Malaria, another type of viral infection that can affect the brain, is rarely seen in the United States, but it is all too common in third world countries where it continues to be a major health problem. Only about 1,200 cases of malaria are diagnosed in the United States each year, and most are in immigrants and travelers returning from areas with a high risk of malaria, such as sub-Saharan Africa and the Indian subcontinent.

This disease is caused by the Plasmodium falciparum (a protozoa) and is transmitted to humans by an infected female Anopheles mosquito. Symptoms of malaria include fever, chills, headache, nausea, and muscle aches. People can feel ill as early as eight days after infection or as long as one year later, though the more typical period is from 10 to 28 days. The mortality rate is high, but those who survive suffer little permanent brain damage.

Malaria continues to be prevalent because of changes in land use, global climatic changes, disintegration of health services, armed conflicts, and mass movements of refugees. The spread of the disease has also been facilitated by the ease of international travel. Perhaps most alarming is that strains of the parasite are becoming resistant to anti-malarial drugs.

African sleeping sickness, which is caused by a trypanosome and transmitted by the tse-tse fly, causes a meningoencephalitis that leads to a febrile (fever) stage and then lethargy. In later stages of the disease, and after a variable incubation period that ranges from weeks to months, there may be tremors, lack of coordination, convulsions, paralysis, mental disturbance, apathy, and drowsiness. Treatment with medication is usually successful.

> **CAUTION**
>
> **Code Blue**
>
> Malaria kills more people than any other communicable disease except tuberculosis. It occurs in more than 100 countries, and the World Health Organization estimates that 300 to 500 million cases occur each year, 90 percent of which are in sub-Saharan Africa. Approximately one million people die of the disease each year, and malaria kills one child every 30 seconds.

AIDS

Diseases related to acquired immunodeficiency syndrome (AIDS) often affect the central nervous system. The brain itself may be directly affected by diseases the body would otherwise be able to fight off, such as toxoplasmosis and cytomegalovirus. The clinical picture may be of an acute, subacute, or chronic meningoencephalitis. A chronic encephalitis giving rise to progressive dementia also occurs with AIDS. Slow deterioration of cognitive and behavioral functions may be slow and insidious. Apathy and flat affect may be present as well. Ultimately, affected people may be unable to speak or control their bladders or bowels.

Words of Wisdom

A **biopsy** is the removal and examination of tissue, cells, or fluids from the body.

In the early 1980s, neurosurgeons often performed brain *biopsies* on patients with AIDS who had neurological symptoms. However, because doctors now know what types of viruses and bacterial infections to anticipate in their AIDS patients, brain biopsies are seldom necessary.

Hereditary Degenerative Diseases

Degenerative diseases involve the irreversible deterioration of cells or organs and thus the impairment of the functions that those cells and organs are meant to perform. A number of degenerative diseases of the central nervous system exist; they are named after the clinicians who first described them. For the most part, these diseases are extremely rare and are beyond the scope of this book.

One more common condition is Huntington's chorea, a condition characterized by involuntary and irregular jerking movements and progressive mental deterioration. Loss of nerve cells in the basal ganglia and cortex is prominent in *histological* studies of patients who die

Words of Wisdom

Histology is the study of the structure of tissues.

with this disease. This inherited disease is thought to be due to a gene defect on chromosome four. Symptoms are rarely seen before age 30. No treatment is available for the dementia, but numerous drugs (similar to those used in Parkinson's disease) are available to treat the movement disorder. Nevertheless, Huntington's chorea is usually fatal.

Congenital Anomalies

Children are sometimes born with brain or spinal cord abnormalities, and, in many cases, the cause is unknown. The following sections describe some of the most common abnormalities.

Hydrocephalus

Hydrocephalus generally refers to a condition characterized by an excess of fluid within the ventricles of the brain. This excess leads to enlargement of the ventricles and subsequently the head of the newborn. Hydrocephalus appears in 3 to 4 babies per 100,000 births.

All of the factors responsible for excess fluid in the newborn are not well known. The condition may be the result of excess production of cerebrospinal fluid, malabsorption of cerebrospinal fluid, or a blockage of the normal pathways of cerebrospinal fluid. Some causes of hydrocephalus are clear, however. These causes include an obstruction of the channel that carries spinal fluid from the third ventricle to the fourth ventricle, tumors, infection, and hemorrhage.

Also, a group of rare congenital lesions can lead to hydrocephalus. For example, Arnold-Chiari malformation is characterized by a downward displacement of the cerebellum, medulla, pons, and fourth ventricle into the upper cervical spine, which causes stretching and narrowing of the aqueduct.

Encephalocele is a rare condition characterized by a disruption of the meningeal covering of the brain that allows brain tissue and cerebrospinal fluid to fill a skin-covered sac, which is usually at the base of the skull in the occipital area.

Treatment of hydrocephalus varies based on its cause. For example, if the ventricular enlargement is caused by blockage that is the result of a tumor, removal of the tumor will usually cure the problem. In most cases, however, a "cure" is not possible, but the hydrocephalus can be controlled with a *shunt*.

Myelomeningocele is characterized by a defect in the spine in combination with a defect in the development of the spinal cord in the first six weeks of intrauterine life and is manifested by the appearance of a sac filled with cerebrospinal fluid and neural tissue anywhere along the spine, but most commonly in the lumbar or lumbosacral area. This unfortunate condition occurs in 1 or 2 babies per 100,000 births.

Myelomeningocele is seen in association with hydrocephalus in a large number of cases. These children may have paralysis of the lower extremities and no control over bowel and bladder function. Treatment involves the surgical removal of the sac with preservation of the neural contents, treatment of the associated hydrocephalus when present, and then a lifelong period of intensive physiotherapy.

Words of Wisdom

A **shunt** is a device put in place during surgery that diverts (shunts) cerebrospinal fluid from the ventricular system into the abdominal cavity where it is absorbed by tissues.

Down's Syndrome

Down's Syndrome is a chromosomal abnormality; people with this syndrome have an extra chromosome. The incidence of Down's Syndrome varies greatly with the age of the mother. In women of early childbearing age (up to about age 25), it is seen in 1 in 2,000 live births, but in women over age 40, the risk rises dramatically to 1 in 100 live births.

Children with Down's Syndrome usually have a characteristic appearance at birth that allows the doctor to diagnose the syndrome in the nursery. The baby's head as a whole tends to be round, but the back of the head is flat. The fontanel (the soft spot on top of the baby's head) is larger than normal, and the shape and position of the baby's ears are abnormal. Often a baby with Down's Syndrome has an unusual gap between the first and second toes and poor muscle tone in the extremities. The baby's mouth usually stays open because of an enlarged protruding tongue. Often the palm of the hand has a single crease, and genitalia tend to be small or underdeveloped.

> **Gray Matter**
>
> A normal cell has 46 chromosomes: 22 pairs plus 2 sex chromosomes, either XX or XY. A female gets one X chromosome from her mother and one X chromosome from her father; a male receives the X chromosome from his mother and the Y chromosome from his father.

Physical and mental development usually is severely retarded in children with Down's Syndrome. Nevertheless, these physically and mentally challenged young people may enjoy long and productive lives with support and physical and occupational therapy.

Amniocentesis and measurement of a substance called alpha-fetoprotein has been an important advance in the intrauterine diagnosis of Down's Syndrome and other congenital neurological anomalies. Genetic counseling has also been helpful to families who have had children with these abnormalities and want to consider further pregnancies.

Brain Attack

Stroke is the third most common cause of death in the United States following heart disease and cancer. Though the conventional wisdom is that stroke primarily afflicts the elderly, statistics show that more than one-fourth of the victims are less than 65 years old.

Stroke is the most common of the disabling neurological diseases; it's more common in men than women and is seen with increasing frequency with advancing age. Risk factors for stroke are hypertension, cigarette smoking, elevated cholesterol, heavy alcohol consumption, use of oral contraceptives, and abuse of cocaine, amphetamines, or heroin. The incidence of stroke in the general population has been decreasing during the past two decades mainly because of the recognition of risk factors and the treatment of hypertension.

Chapter 4 noted that the U.S. government certified the 1990s as the "Decade of the Brain." This designation led to an increase in interest in disorders of the brain and in particular an interest in research of stroke.

A major impact on diagnosis and treatment of stroke came with the designation of stroke as a "brain attack." This appellation was made so that the condition would be given the same importance as a heart attack and patients, family, and physicians would recognize the same urgency for the need for emergency treatment.

According the American Stroke Association, the following are warning signs of stroke:

- Numbness or weakness on one side of the body (can't move one arm) or face (mouth droops on one side)
- Inability to understand words or talk clearly
- Poor vision in one or both eyes
- Inability to walk well, dizziness, or falling
- Very bad headache

If you or someone in your family is experiencing the symptoms of a brain attack, call 911 immediately.

> ### IQ Points
>
> Each year, about 600,000 people suffer a stroke. About 500,000 of these are first attacks, and 100,000 are recurrent attacks. About 1 in 15 people die of stroke, more than 150,000 annually. On average, someone in the United States suffers a stroke every 53 seconds; every 3.3 minutes someone dies of a stroke. More women than men die from strokes, and nearly half of stroke deaths occur outside hospitals.

Types of Brain Attacks

A stroke is characterized by the acute onset of a neurological deficit that lasts more than 24 hours and is the result of a disturbance of brain circulation. The specific symptoms of a stroke are related to the area of the brain affected by the circulatory deficit. The neurologic deficit may be catastrophic and immediate or may occur over hours or even days with relentless progression (unless there is emergency treatment). A neurologic deficit that is progressive over weeks or months is not the result of a stroke and is more likely related to a tumor or degenerative disease. Strokes are generally of two types: Ischemia and hemorrhage.

Ischemia is a condition in which an area of the brain is deprived of blood flow and therefore the area is deprived of oxygen. This deprivation will lead to neuronal cell death if the blood flow is not promptly restored to the area. If brain cell death is extensive, then a stroke (or cerebral infarction) has occurred.

This oxygen deprivation can happen in a number of ways. A thrombosis (blood clot) may block a blood vessel, or an embolus (a blood clot that arises in another part of the body, usually the heart) may do the same thing, resulting in ischemia to a part of the brain and ultimately infarction. In most cases, atherosclerosis (having fatty deposits and deposits of calcium or hardening) of major blood vessels in the neck, at the base of the brain, or within the brain itself is the cause of ischemia.

Eighty percent of strokes are ischemic. The remainder are hemorrhagic. A hemorrhagic stroke is characterized by arterial bleeding that may occur in any area of the brain. Bleeding into the substance of the brain is known as intracerebral hemorrhage, which causes destruction or compression of brain tissue and subsequent cell death. A ruptured artery around the brain is called a subarachnoid hemorrhage.

Two important forms of stroke, sometimes referred to as "mini-strokes," are a RIND (reversible ischemic neurological deficit) and a TIA (transient ischemic attack). The RIND is characterized by a deficit lasting more than 24 hours that clears in a few days. RINDs may be recurrent. TIA refers to an episode of focal neurological symptoms (that is, involving one specific area of the brain) lasting up to 30 minutes with complete resolution within 24 hours. About 30 percent of individuals with recurrent TIAs go on to have a major stroke. These episodes are usually caused by a thrombosis or embolism.

For instance, a 64-year-old male saw his family physician because of an episode of blindness in the left eye that lasted for about five minutes. This was followed by blurred vision, which cleared completely in about ten minutes. The patient reported that three months earlier he had numbness in his right arm and leg, accompanied by slurred speech that lasted about 20 minutes before going away. His physician told him that these episodes sounded like TIAs (transient ischemic attack). He ordered an MRI scan which was reported as normal. A neurological consultation was requested. A carotid Doppler study (a type of ultrasound) showed a decrease in blood flow in the left carotid artery in the neck. An arteriogram showed a 90 percent stenosis (constriction) of the left internal carotid artery in the neck. A left carotid endarterectomy was performed to open the artery and remove atherosclerotic plaque. The patient made a full recovery and was started on aspirin. He had no further TIAs.

Though stroke is typically associated with older people, and they are most likely to be affected, brain attacks can affect people of all ages, including babies prior to birth. Some victims have inherited weaknesses in the walls of the arteries in their brains and that makes them more susceptible to strokes.

Treatment for Brain Attacks

Treatment of TIAs are important in the prevention of a major stroke, and antiplatelet drugs are currently the treatment of choice. One recommended drug is aspirin taken in a dosage of 81 milligrams per day. (Be sure to consult with your doctor before beginning any medication program.)

If an obstruction of the carotid artery in the neck can be demonstrated in a patient having TIAs, surgery (called carotid endarterectomy) may be advised, although a number of studies suggest that the long-term results are the same with antiplatelet drugs as with surgery. Anticoagulant drugs are also a mainstay of treatment in this situation.

The signs and symptoms of both ischemia and hemorrhage strokes are the same and are directly related to the specific area of the brain involved. For the purpose of treatment, differentiating one from the other is extremely important. The neurological examination and a carefully taken history from the patient is essential in making a diagnosis. An emergency CT scan is very helpful in the diagnosis of hemorrhage. An MRI scan is not as helpful. The MRI scan, however, is more accurate in assessment of the extent of infarction after bleeding has been ruled out.

The treatment for hemorrhagic stroke is in general symptomatic unless there is evidence of increased intracranial pressure and the hemorrhage (blood clot) is in a non-eloquent area (an area that can be operated on without killing the patient or causing severe neurological deficit) of the brain.

In the patient with a major stroke, treatment should be instituted within four hours of the brain attack if at all possible. Emergency treatment with drugs such as TPA (tissue plasminogen activator) may dissolve the clot and lead to a considerably better outcome.

The major risk or complication of TPA treatment is hemorrhage, so patients undergoing this treatment require intensive care with close monitoring. Unfortunately, not all hospitals are equipped to provide this emergency treatment.

Brain Death

Most people believe death occurs when a person's heartbeat and breathing stop. That is not legally the case. After all, as you probably know from watching medical dramas on TV, machines can keep a person alive by breathing for them and keeping the heart beating. They will look like someone who is alive but asleep.

Death is now typically defined on the basis of brain function. "Brain death" is defined as irreversible cessation of all brain function. It is not as easy to determine as you might think and, obviously, this isn't something you want to make a mistake about. In fact, the criteria were established by the President's Commission for the Study of Critical Problems in Medicine and Biomedical and Behavioral Research in 1981:

♦ The patient must be unresponsive to sensory input including pain and speech.

♦ Brainstem reflexes must be absent, including the *"Doll's eye" reflex*.

♦ Pupillary reflexes are absent and eye movement does not occur after placing cold water in the ear.

♦ There is no respiration.

♦ The cause of coma is known and irreversible.

♦ There is no evidence of drug intoxication, hypothermia (body temperature below 90 degrees F), shock, or neuromuscular blockade (anesthetic).

♦ All of the above must be present for six hours and the EEG is flat.

♦ All of the above must be present for 12 hours and there is no EEG available.

♦ All of the above must be present for 24 hours for an injury where the brain is denied oxygen without an EEG.

Words of Wisdom

The "Doll's eye" reflex refers to the way the eyes of a doll move in the opposite direction from where the head is turned.

Organs for Sale or Rent

There is an acute shortage of organs for transplantation. Many seriously ill people are anxiously hoping for someone to donate an organ, with hearts, kidneys, and livers in particular demand. The two biggest problems are that not enough people are willing to donate their organs and that those who may be willing have illnesses that rule them out as candidates. For example, you can't take the liver from an alcoholic or a heart from someone who had a heart attack, or a kidney from someone with kidney disease. A preexisting illness will disqualify a donor, but only if it affects the organ to be transplanted. For example, a person who dies of a heart attack can't donate their heart, but if their liver is perfectly healthy it could be used in a transplant.

The most common sources for transplant organs are relatively young individuals who have sustained lethal head injuries. Youth matters because it is more likely that before the injury they were healthy.

In many states you can have a notation on your driver's license that you agree to be an organ donor. It is important for the family of a person who suffers a head injury to know the medical history of that individual to help doctors determine if their organs are healthy enough to be used. For example, it is a difficult and sensitive task, but a doctor will ask the family if the person with the head injury was healthy or if they were chronic alcoholics, or if they ever had kidney or lung disease. As you can see people who abuse their bodies with drugs and alcohol may not only die themselves, but deny others the chance to live.

The Least You Need to Know

- A significant number of head injuries could be avoided if everyone wore seatbelts in cars and helmets on bicycles and motorcycles.
- More than half of all brain tumors are malignant, but a large number of tumors can be cured.
- Viruses and infections can affect the brain and, in the case of diseases such as meningitis, rabies, and malaria, can cause catastrophic illness and even death.
- "Brain attack" or stroke is the third most common cause of death, and its symptoms should be treated with the same urgency as those for a heart attack.

Chapter 18

Headaches and Neurological Disorders

In This Chapter

- ◆ There are headaches, and then there are headaches
- ◆ Celebrities raise the profile of Parkinson's disease
- ◆ Seizures share their secrets
- ◆ A president loses his memory
- ◆ Mad cows and humans confront the same disease

This chapter describes the more common maladies affecting the brain. These conditions are painful and keep people home from work, but they don't usually put people in the hospital.

This chapter also describes several conditions that have a more serious long-term impact. They eat away at people's bodies and minds and, unfortunately, are not yet curable.

Your Aching Head

Headache may be the most common pain of all. (Of course, "I have a headache" is one of the most common excuses as well.) In the United States, 9 out of 10 people have at least one headache a year. Approximately 40 million Americans suffer from chronic headaches.

Chapter 17 listed headache as an important symptom of many brain diseases, including stroke. This section discusses headache as its own malady unrelated to any other specific disease. Headaches can be divided into a number of categories. The three most common are migraine, cluster, and tension headaches.

Migraine Headaches

An estimated 28 million people suffer migraines in the United States. Many patients get their first migraine during their teen years. Migraines also appear to run in families; children have a 50 to 60 percent chance of suffering them if one parent does. For some unknown reason, women, particularly those younger than 40, are much more susceptible to them than are men; in fact, 75 percent of the total number of people who experience migraines are women.

Furthermore, people rarely get a single migraine; migraines are usually a recurring problem for those who suffer from them. When you get a migraine you feel as though your head is about to explode. Migraine headaches are characterized by nausea, vomiting, and photophobia (a fear of light—not photographs!) in conjunction with a severe headache on one side of the head (unilateral). Visual, auditory, or gastrointestinal symptoms may precede the onset of the headache. The most common symptom that precedes classic migraine headaches involves a problem with vision, for example, a sudden loss of some vision. In the majority of patients, a migraine lasts more than 2 hours, but less than 24.

Scientists believe that a narrowing of blood vessels (vasoconstriction) inside the brain causes the symptoms before the migraine hits and the enlargement of the blood vessels (vasodilation) on the scalp and face cause the migraine itself. Migraine attacks may be precipitated by a number of things. Foods, such as certain cheeses, hot dogs with nitrite preservatives, and chocolate with phenylethylamine, and certain food additives such as monosodium glutamate can all trigger migraines. Other causes include menstruation, contraceptives, and bright lights.

Treatment of the acute migraine attack involves the use of analgesics starting with aspirin, acetaminophen (Tylenol), or ibuprofen (Motrin, Advil). Ergotamine, a powerful vasodilator, should be given at the same time. If the acute attack does not respond, stronger narcotics such as meperidine (Demerol) may be necessary.

One of three drugs may be helpful in preventing migraine attacks. They are propanolol, amitryptyline, and valproic acid. Doctors may have a patient try one or more of these drugs to see what works to prevent the migraine attacks. Several other drugs are also available for treatment; unfortunately, they don't always work. The decision regarding the drug of choice is dependent upon the clinical condition of the individual and whether the individual also has such conditions as heart disease, asthma, or diabetes. Consult your doctor for treatment if you suffer from migraines.

Cluster Headaches

You're sitting in front of the TV and suddenly you feel an excruciating pain around your eye and your nose starts to run. This may be an indication of a cluster headache.

Most individuals who have cluster headaches feel that they are the most painful of all the various headaches. These severe unilateral headaches occur in bunches and strike men almost 10 times as often as women. Although a migraine headache may be bilateral (affecting both sides of the head), a cluster headache is always unilateral. The headache may occur at night and daily for days or weeks. Once the cluster headache clears, it may not surface again for months or years.

Cluster headaches are usually characterized by severe pain around the eye, which may be accompanied by redness, tearing, nasal stuffiness, runny nose, and a burning sensation in or behind the eye. An associated Horner's syndrome (remember meiosis, ptosis, and anhydrosis from Chapter 12?) may also occur. The headache may last from a few minutes to a few hours. Patients with cluster headaches rarely have a family history of headache.

Treatment of cluster headaches is similar to the treatment of migraines; however, Prednisone has been shown to have a dramatic effect in relieving the headaches. Indomethacin has also been used with good results. Consult your doctor for treatment if you suffer from a cluster headache.

Tension Headaches

The tension headache is probably the most common type and is one with which we are all familiar. This headache is often bilateral and often occurs in the occipital area but may involve the frontal area or involve the entire head. It is often described as feeling as though a vise is tightening around the head, but it is almost never described as throbbing. Tension headaches are seen in men and women alike, but they rarely occur before age 21.

The symptoms of tension headaches include tightness of the neck and scalp muscles. These headaches are not associated with nausea, vomiting, or any type of early symptoms. A tension headache usually occurs everyday and may be present on awakening in the morning as well as at bedtime.

Before a doctor starts treatment for a headache, he or she should take a careful history from the patient. Having an understanding, reassuring doctor who accepts the pain as real, not imaginary, helps greatly. Patients should see their doctors often and be encouraged to discuss their emotional problems.

Headache patients often need reassurance that the headache has no serious organic cause. When in the throes of a serious or recurring headache, some of us are quick to think the worst, that perhaps we have a brain tumor. The truth is that the vast majority of headaches are either a migraine, cluster, or tension headache and only a very small percentage are caused by a brain tumor.

In some cases, the doctor may want to perform certain diagnostic studies such as an MRI scan, although this scan is done in a very small number of cases because doctors diagnose a tension headache on the basis of the history given by the patient and the physical examination. Most people with tension headaches benefit from taking mild analgesics, such as aspirin or acetaminophen (Tylenol). Physical therapy, psychotherapy, and relaxation techniques, including feedback, may be helpful in rare cases. Removing irritants and stresses before resuming or continuing a fully active life may be the most important aspect of tension headache treatment.

Trigeminal Neuralgia

People who suffer from trigeminal neuralgia (tic douloureux) may not know anything is wrong until they brush their teeth and feel a stabbing pain shoot through their jaw. This painful condition is characterized by a lancinating (tearing, darting, shock-like) pain in the distribution of the trigeminal nerve (V). This pain may be in the lower jaw (the mandibular area), adjacent to the nose and in front of the ear (the maxillary area), or around the eye or forehead (the ophthalmic area). The pain is often described as a lightning-like excruciating pain and is often set off by touching a trigger point on the face or by performing an activity such as chewing or brushing the teeth.

The cause of trigeminal neuralgia is not known, although in many cases an abnormal blood vessel has been found compressing the trigeminal nerve where it enters the brain stem. Treatment with Tegretol and, in some cases, Dilantin has been effective. In cases resistant to medical therapy, surgical treatment has been used. The surgical procedure involves making an opening into the cranium at the base of the skull and then isolating the trigeminal nerve where it enters the brain stem. The surgeon then removes the pressure on the nerve being exerted by a blood vessel by moving the vessel away from the nerve.

Cerebral Palsy

Cerebral palsy (CP) is not a specific disease, but rather a descriptive term used to denote a group of disorders characterized by damage to the nervous system in utero, at birth, or just after birth. CP has also sometimes been called "congenital brain syndrome" and usually refers to the condition of children who have a static (fixed and unchanging) motor deficit.

The condition may be caused by a lack of sufficient oxygen in utero, at birth, or just after birth; an infection, malformation; or a hemorrhage in the brain. Respiratory problems or congestive heart failure related to generalized infection may cause decreased oxygen in the blood (hypoxia) leading to a decrease in the blood supply to a portion of the brain (ischemia) and may explain what is then seen as the signs of cerebral palsy. Trauma during delivery may also play a role in causing the condition.

In addition, an increased risk of cerebral palsy exists when a baby is born weighing less than 4.5 pounds (2 kilograms) and the five-minute *Apgar score* is less than three. Fetal monitoring and intensive neonatal care have substantially decreased the number of children with cerebral palsy.

Children with cerebral palsy present many clinical problems as they enter early childhood. Most are mentally and physically challenged. Psychological testing is important early in life. Neurological evaluation and follow up for many years is beneficial for the child and the family. Orthopedic evaluation and treatment and intensive physiotherapy is also extremely helpful, especially as the child gets older.

> **Words of Wisdom**
>
> The **Apgar score** refers to the evaluation of newborns based on five characteristics: complexion, heart rate, response to stimulation of the sole of the foot, muscle tone, and respiration. Infants are rated 0, 1, or 2 for each characteristic, with a total of 10 being a perfect score. For mothers who just gave birth, this score is the most important grade her child will ever get.

Parkinson's Disease

Most of us don't know much about Parkinson's disease, but we've now become familiar with the symptoms because of the number of well-known individuals who in recent years have publicly admitted that they have the malady. Parkinson's disease occurs in about one to two people out of every thousand. It is found equally in both genders and becomes increasingly common with increasing age.

Parkinson's disease of unknown cause is the most common type. However, Parkinson's disease may be caused by a number of factors. An epidemic of viral encephalitis that occurred between 1918 and 1923 is thought to be responsible for many cases of Parkinson's disease that were seen decades later. Therapeutic drugs such as phenothiazines (a urinary antiseptic) or toxic substances such as carbon monoxide also can cause Parkinson's disease.

The disease is characterized by a variety of signs, the most prominent of which are tremor, mask-like face, dysarthria (difficulty speaking), stooped posture, rigidity of muscles, abnormalities of gait, slowness, and lack of movement without cognitive decline. Though the effects of the disease are primarily physical rather than mental, a decrease in cognitive abilities and even depression may occur late in the course of the disease as the person tries to cope with the debilitating nature of the illness.

A great deal of research shows that those with Parkinson's disease have a chemical imbalance between the neurotransmitters dopamine and acetylcholine caused by a marked decrease in dopamine. There is also a decrease in the neurotransmitter norepinepherine. Microscopic examination of the brain of a patient with Parkinson's disease shows a decrease in pigmentation and neuronal loss in the substantia nigra (a tiny area about the size of a quarter that produces most of the dopamine used in the brain) and other brainstem structures and the basal ganglia.

IQ Points

Parkinson's disease did not get a great deal of public attention until three celebrities became afflicted with it. Perhaps the most famous person in the world, boxer Muhammad Ali, has the most serious visible case, with severe tremors and slurred speech. Former U.S. Attorney General Janet Reno also disclosed during her term that she has the disease. Popular actor Michael J. Fox, the youngest of the three, gave up a starring role on television to focus full-time on raising awareness about the disease and money for research.

At one time, doctors thought the removal of the cerebellum would cure the condition, but this procedure did not work because it was later learned that the disease was not a result of faulty cerebellar function. Surgical treatment of Parkinson's disease for many years involved making lesions in the part of the brain thought to be responsible for the causation of tremors. Recent surgical procedures have utilized the implantation of a high-frequency stimulating electrode, which is placed in the thalamus to relieve the tremors. A battery pack is placed under the skin of the chest and attached to the brain stimulator. The patient can then turn the brain-stimulating electrode on and off at will. The battery can thus be turned off at night when stimulation is not needed.

Treatment for Parkinson's disease has seen dramatic improvement over the past few years. Today, medical treatment is directed toward restoring the dopaminergic-cholinergic balance by blocking the effect of acetylcholine.

Dudley's Disease

Progressive supranuclear palsy is a degenerative disease of unknown cause. This unfortunate condition was brought into the public eye by the death of actor Dudley Moore, the star of movies such as *Ten* and *Arthur*. The disease affects the subcortical gray matter, which causes a failure of vertical and then horizontal gaze, so that the affected individual has a difficult time focusing on an object in front of him. There is marked postural instability so the individual tends to fall. The patient has facial weakness, difficulty with speech and swallowing, and becomes demented. He or she may have exaggerated and emotional responses, such as laughing or crying when neither is appropriate. This condition affects men twice as often as women. Death usually comes from four to seven years following the diagnosis.

Epilepsy

The word *epilepsy* comes from the Greek word for seizure; therefore, it is appropriate to spend a few moments here discussing seizures. A *seizure* is a sudden attack or convulsion due to an uncontrolled burst of electrical activity in the brain that can result in a variety of uncontrollable actions, such as shaking, twitching, and losing consciousness. Most of us have seen people in the throes of seizures on television, if not in real life, and it can be a frightening sight.

Seizures may be classified as generalized or partial seizures. Generalized seizures occur when the electrical disturbance flows through the whole brain at once. Partial seizures are those that occur when the electrical activity is centered in one part of the brain and affects the physical or mental activity that that particular area controls. Sometimes the seizure can begin in one area of the brain and spread to the rest of the brain.

Generalized seizures are further classified as tonic-clonic (grand mal), absence (petit mal), or myoclonic (spasm with rigidity and relaxation). Tonic-clonic seizures usually involve a loss of consciousness. As the individual regains consciousness, he or she is confused and has a headache. In a condition known as status epilepticus, the seizures fail to stop spontaneously. This condition presents a medical emergency because permanent brain damage may occur if the condition goes untreated.

Partial seizures may be simple partial seizures or complex partial seizures. People suffering partial seizures remain awake and aware and can sometimes talk normally as they are experiencing them. Simple partial seizures may begin with motor, sensory, or autonomic symptoms and may include déjà vu (the sensation that a new experience is being repeated). Uncontrolled movements can occur in just about any part of the body and sometimes begin with the shaking of a hand or foot that spreads to more of the extremity or the whole side of the body. Depending on the part of the brain affected, the seizure may cause a variety of sensations and emotions, including feeling fearful, hearing voices that aren't there, smelling strange smells, laughing or crying uncontrollably, and experiencing distorted perceptions of the way things look. There may be an *aura* prior to the onset of the seizure.

Complex partial seizures, previously called temporal lobe seizures, are characterized by impaired consciousness, without loss of consciousness. People don't know what they are doing and can't control their movements, speech, or actions. Automatism, a form of involuntary motor movements, may occur with these seizures.

Words of Wisdom

An **aura** is a subjective phenomenon or sensation marking the onset of a seizure. This sensation may be auditory (hearing strange sounds), visual (seeing strange things), motor (moving strangely), or gastrointestinal (feeling abdominal pain).

Most seizures don't last long and end by themselves after a minute or two. In some cases, however, they may be more serious. Seizures that occur throughout the day and fail to respond to drug therapy may indicate a need for surgical therapy. If diagnostic evaluation, including EEG, MRI, and PET scans, can pinpoint the source of seizure activity within the brain, this area can be removed. Unilateral temporal lobe excision is employed to cure many seizure cases, and in extremely rare cases, hemispherectomy (removal of one half of the brain!) has been employed to treat uncontrollable seizures that occur all day long.

What causes seizures? Many causes of seizures have been identified, including the following:

◆ Head injury
◆ Benign febrile convulsions during childhood
◆ Stroke
◆ Tumors
◆ Meningitis or encephalitis
◆ Hypoglycemia
◆ Hyponatremia (a decreased sodium content in the bloodstream)
◆ Drug overdose
◆ Drug withdrawal
◆ Eclampsia (in pregnancy associated with hypertension, protein in the urine, and swelling)
◆ Hyperthermia (body temperature above 106 degrees)

Code Blue

People having seizures cannot swallow their tongues, so there is no need to force their mouths open, but they may bite their tongue because of clenched teeth so a firm object such as a spoon or a tongue blade should be placed between the teeth.

The treatment of seizures related to these conditions is dependent upon recognition of the underlying cause of the seizure and therapy directed at this cause.

About 0.3 percent of the general population suffers from epilepsy of unknown cause (idiopathic). Among the famous people who are believed to have suffered from epilepsy or experienced seizures were Alexander the Great, Julius Caesar, Napoleon, Beethoven, and Agatha Christie. Epilepsy is not contagious. The condition does not affect intelligence, but it may affect a child's ability to learn. Children who have epilepsy sometimes grow out of the condition.

The treatment of idiopathic epilepsy is well documented and involves finding the right combination of medications for each individual sufferer. Epilepsy is not a mental illness, though people who have mental illnesses may also have seizures. To date, research has not conclusively demonstrated that typical seizures lasting a few minutes or less cause permanent damage to the brain; however, status epilepticus (seizures that are continuous, lasting for more than several minutes) may cause permanent brain damage.

Code Blue

If someone has a generalized tonic-clonic (grand mal) seizure, the Epilepsy Foundation recommends taking the following measures:

- Keep calm and reassure other people who may be nearby.
- Don't hold the person down or try to stop his or her movements.
- Time the seizure with your watch.
- Clear the area around the person of anything hard or sharp.
- Loosen ties or anything around the neck that may make breathing difficult.
- Put something flat and soft, like a folded jacket, under the person's head.
- Turn the person gently onto one side to help keep his or her airway clear. Do not try to force the mouth open with any hard implement or with fingers.
- Don't attempt artificial respiration except in the unlikely event that the person does not start breathing again after the seizure has stopped.
- Stay with the person until the seizure ends naturally.
- Be friendly and reassuring as the person's consciousness returns.
- Offer to call a taxi, friend, or relative to help the person get home if he or she seems confused or unable to get home.

Multiple Sclerosis

Multiple sclerosis or MS is a very common neurological illness. The age of onset is between 20 and 40, and it is much more typical in women than men. The pathological feature of the disease is extensive demyelination (remember that myelin is the fatty substance that envelops the axon or nerve fiber of the neuron) in scattered areas of the white matter of the brain and spinal cord, as well as the optic nerve. Subsequent scarring (gliosis) in the area gives rise to the *plaques* that are seen on imaging studies and at autopsy. The damage to the insulation of the neurons can block nerve signals and cause them to become garbled and can slow down the sending of the messages.

Although the cause of the disease is unknown, MS is currently thought to be an autoimmune disease, which means that the affected person's own immune system is destroying the myelin. Because the disease affects all parts of the central nervous system, the early symptoms of the disease can vary, but they most often include weakness, numbness, tingling, sudden loss of vision or blurring of vision in one eye, and urinary urgency or frequency. These symptoms may last days or weeks and then clear completely only to recur again at a future date.

The diagnosis of MS isn't easy because the individual symptoms are common to several other medical problems. Doctors usually must follow patients closely and monitor the different neurological signs they display over time. For example, in one episode a patient may complain of numbness or tingling in the face or an extremity; weeks or months later, a second episode could involve blurred vision; and weeks or months later, yet another episode may involve bladder dysfunction or weakness in an extremity. Patients with MS may have abnormalities in their cerebrospinal fluid, and imaging studies may show the presence of plaques.

Treatment of MS is the basis of ongoing research in many centers. Studies show that medication is quite often helpful in the reduction of the rate of relapses and the slowing of the progression of the plaques, and steroids may hasten recovery from an acute attack. However, there is no known cure for MS at this time.

Alzheimer's Disease

For many years little understood or publicized, Alzheimer's disease has received a great deal of attention since President Ronald Reagan let the world know that he suffered from this chronic debilitating condition. On November 5, 1994, President Reagan published the following letter:

My fellow Americans,

I have recently been told that I am one of the millions of Americans who will be afflicted with Alzheimer's disease.

Upon learning this news, Nancy and I had to decide whether as private citizens we would keep this a private matter or whether we would make this news known in a public way. In the past Nancy suffered from breast cancer, and I had my cancer surgeries. We found through our open disclosures we were able to raise public awareness. We were happy that as a result, many more people underwent testing. They were treated in early stages and able to return to normal, healthy lives.

So now we feel it is important to share it with you. In opening our hearts, we hope this might promote greater awareness of this condition. Perhaps it will encourage a clearer understanding of the individuals and families who are afflicted by it.

At the moment I feel just fine. I intend to live the remainder of the years God gives me on this Earth doing the things I have always done. I will continue to share life's journey with my beloved Nancy and my family. I plan to enjoy the great outdoors and stay in touch with my friends and supporters.

Unfortunately, as Alzheimer's disease progresses, the family often bears a heavy burden. I only wish there was some way I could spare Nancy this painful experience. When the time comes, I am confident with your help she will face it with faith and courage.

In closing, let me thank you, the American people, for giving me the great honor of allowing me to serve as your President. When the Lord calls me home, whenever that day may be, I will leave with the greatest love for this country of ours and eternal optimism for its future.

I now begin the journey that will lead me into the sunset of my life. I know for America there will always be a bright dawn ahead.

Thank you, my friends. May God always bless you.

Sincerely,

Ronald Reagan

This sincere outpouring of emotion by the former president moved many people, and his revelation undoubtedly gave hope to many of the four to five million people in the United States who suffer from this condition.

Alzheimer's disease is characterized by progressive mental deterioration. Work and social interaction becomes increasingly difficult as thought processes and memory deteriorates. There may be apathy and loss of spontaneity. Facetiousness and a disturbance in the formation of moral and ethical judgments may be seen as well. The person also may be incapable of handling business and financial affairs. The rate of downhill progression is extremely variable and may take from several months to many years. Eventually the affected person is unable to comprehend what is going on around him or her and may have difficulty recognizing and communicating with family members.

IQ Points

In January 2002, researchers at UCLA announced they had created the first test that records the onset of Alzheimer's disease. This test could be an important development because by the time symptoms for Alzheimer's appear, patients already have considerable brain damage. Current diagnostic tests are only about 55 percent accurate, but the new procedure, which involves a PET scan and the injection of a chemical tracer, may improve early diagnosis and lead to more effective treatment.

Some people with a family history of Alzheimer's disease may be at greater risk of suffering from the condition in later life, but the cause of the disease is unknown. The changes that occur in the brain, however, are well documented. It appears that neurons are destroyed when small hollow fibers displace normal cellular structures (a process called neurofibrillary degeneration); at the same time, plaques accumulate in the cerebral cortex.

Although Alzheimer's usually does not affect people until they are in their 60s or 70s (1 in every 10 people over 65 is affected and nearly half of those over 85 are affected), earlier onset is not unusual. The younger people are when they get the disease, the worse the prognosis is. The disease also affects more women than men.

There is no known cure for Alzheimer's disease, but in recent years, a number of medications have become available that are thought to slow the progression of the disease and even ameliorate some of the severe memory loss.

Normal Pressure Hydrocephalus

Normal pressure hydrocephalus is a condition seen in elderly individuals that is characterized by a triad of symptoms: dementia, gait disturbance (trouble walking), and urinary incontinence. Imaging studies (CT scan or MRI) demonstrate an enlargement of the ventricles of the brain. However, there is no evidence of headache, nausea, vomiting or other symptoms of increased intracranial pressure. Spinal taps reveal normal pressure. The onset of *dementia* from the disorder is insidious and must be differentiated from other neurological disorders that may at first appear similar.

Surgical treatment involves the shunting of cerebrospinal fluid (CSF) either from the ventricle of the brain or the lumbar CSF (from the area where CSF is removed when doing a spinal tap) into the abdominal cavity. However, the procedure is not always successful in relieving the symptoms. When the procedure is successful, the changes in the patient are dramatic.

Words of Wisdom

Dementia involves the deterioration of mental functions, such as memory, concentration, and judgment, resulting from an organic disease or a disorder of the brain. It is sometimes accompanied by emotional disturbance and personality changes and typically interferes with a person's daily functioning.

Mad Cow Disease

The publicity given to the infection of British cattle suffering from Mad Cow disease has created widespread fears that humans may be in danger of going crazy, or at least getting sick from exposure to infected animals or eating tainted beef. It was enough to make even the most carnivorous among us go vegetarian. It should come as no surprise that the truth about the disease is quite different from the hype.

Mad Cow disease, which gets its name from the fact that it affects the brain of the infected cow, is a variant of Creutzfeldt-Jakob disease, an invariably fatal transmissible disease of the central nervous system characterized by a rapidly progressive dementia and involving the entire brain and spinal cord. A protein-related infective agent known as a prion is thought to be the cause of this disease.

The clinical picture is one of rapid deterioration over a period of months and includes twitching, rigidity and tremor with abnormal movements, and a loss of the ability to coordinate voluntary muscle movements (ataxia). The disease affects about one in one million people in the United States. *Extremely rare* cases of person-to-person transmission have been attributed to corneal transplantation and cortical electrode implantation. No treatment is available for Creutzfeldt-Jakob disease, and death usually occurs within one year.

Gray Matter

According to the CDC, the risk of getting Creutzfeldt-Jakob disease from eating beef products produced from cattle in Europe cannot be precisely determined, but it appears to be extremely small, perhaps about 1 case per 10 billion servings. Mad Cow disease has not been detected in the United States, and it is extremely unlikely that it will be a food-borne hazard here. In April 2002, the first case of someone in the United States getting sick from infected beef was reported. A British woman living in Florida apparently caught the disease after eating beef in Britain. More than 90 people in Europe have died from eating contaminated beef.

The disease seen in cattle, known to scientists by the intimidating name of bovine spongiform encephalopathy (BSE), is caused by a self-replicating protein. It was first observed in Great Britain in 1984 and has affected more than 200,000 cattle in the last 15 years. When seen in humans, it occurs in young individuals, has a more prolonged course, and is characterized by early psychiatric abnormalities such as depression and personality changes. It is extremely rare for animal-to-human transmission to occur, and no cases have been reported in the United States.

Tics and Tones

The syndrome described by Gilles de la Tourette has become increasingly well-known in recent years because of sometimes negative depictions on popular television shows. About 100,000-200,000 people have Tourette syndrome (TS), with most suffering symptoms before age 18 and with males three to four times more likely to be affected than females.

Symptoms range in severity, with most people having milder ones. Those seen early in the disease are various tics (a spasmodic movement or twitching) of the face, sniffing, blinking, and eye closure. Tics decrease in frequency and intensity during sleep. Sufferers do have some control over these tics so it is not true to say they are involuntary; however, efforts to suppress the tics can lead to a buildup that results in the eventual release of a burst of tics.

Later, as the disease progresses, involuntary vocal tics occur, such as grunts, barks, throat clearing, and verbal utterances that are vulgar or obscene. The extreme instances of obscene verbal tics actually occur in only about 15 percent of the TS cases. There may be self-mutilation, such as severe nail biting, hair pulling, and biting the tongue or lips. Attention hyperactivity disorder and obsessive compulsive disorder are often seen (see Chapter 20 for more on these disorders), but TS does not affect a person's intelligence.

IQ Points
Among the famous people who have suffered from Tourette syndrome are the famous British writer and lexicographer Samuel Johnson, baseball player Jim Eisenreich, and NBA basketball star Mahmoud Abdul-Rauf.

The majority of people with TS are not significantly disabled by their tics or behavioral symptoms, and therefore do not require medication. For those sufferers with more severe cases, medication is available. Medical treatment to decrease the severity and frequency of the tics is successful in about 50 percent of the patients. The symptoms generally decline in severity after puberty and in about 20-30 percent of the cases, the symptoms disappear entirely when the person reaches their twenties.

The cause of TS remains unknown, but recent research suggests the disorder is related to the abnormal metabolism of the neurotransmitter dopamine and perhaps others such as serotonin. A person with TS has about a 50 percent chance of passing the gene on to their children. The child, however, may not develop the same symptoms as the parent; for example, they may have mild tics or none at all.

Facing the Fear

The idea of something going wrong with your brain has to be about as frightening as anything we can imagine, even in our worst nightmares. In practice, most of us have a general fear of brain disease because of the belief that any brain disease will lead to a loss of our mental abilities and may be life threatening.

Think about it from the doctor's point of view for a second. Imagine how difficult it would be to tell someone that there is a problem with their brain. The diagnosis is devastating for both the patient and their family, but the doctor will explain that not all brain disease leads to mental deterioration. For example, most people with Parkinson's disease are awake, alert, aware of their surroundings, and are not intellectually impaired in any way (as evidenced by former attorney general Janet Reno's campaign to become governor of Florida).

Even in many cases where an illness does lead to some impairment, the deterioration in mental function may be gradual over a long period that can allow the patient and family to adjust. For example, the onset of Alzheimer's may force the patient to give up driving a car and then later they may be unable to leave the house.

And keep in mind that new breakthroughs occur in medical science on a daily basis. These include new technologies, improved skills of the medical team, and new treatments. Some ailments can be cured completely. Many tumors can be removed, damaged parts of the brain can sometimes be repaired or the functions taken over by undamaged areas, and patients can often live long productive lives with physical and mental disorders.

The Least You Need to Know

 ♦ Headaches are the most common pain that people suffer. Most people experience tension headaches, but many people have even more severe pains from migraines and cluster headaches.

 ♦ Celebrities such as Michael J. Fox have helped make people aware of the debilitating nature of Parkinson's disease and the importance of research to ameliorate the symptoms and find the cause.

 ♦ Abnormal electrical activity in the brain can trigger seizures. Seizures can range in severity, but they are usually not life-threatening, which means people with epilepsy can live normal lives.

 ♦ Ronald Reagan's admission that he suffers from Alzheimer's disease has helped to make finding a cure and treatments for this horrible, debilitating mental condition a national priority.

Chapter 19

Drugs and Demons

In This Chapter

- ◆ Drinking and smoking
- ◆ Snorting and shooting up
- ◆ Tripping
- ◆ Popping pills

Many illicit drugs, as well as legal medications, mimic the activities of natural chemicals in the brain. They alter communication in the brain because they resemble the normal chemical messages produced by the brain and either block the action of a neurotransmitter or create extra amounts of it. Some of these drugs can be very beneficial; others are extremely dangerous. Sometimes the line between the two extremes is a fine one, and in others it is quite clear.

Abusing any drug can be hazardous to your health, and several of the drugs that are discussed in this chapter can be particularly damaging to the brain.

Some drugs cause changes in the brain that persist even after the user has stopped taking the drug. These changes can produce addiction. The pleasure derived from drugs also implies that addiction has a learned component as well; that is, the user consciously recognizes the connection between their euphoric feeling and the drug.

Not all drugs are addictive, and drugs that are addictive may not immediately hook a user. On the other hand, addiction can be so powerful that just seeing a picture of the place where the user bought the drug may produce a strong craving for it. Different people have varying tolerances to drugs, and this tolerance is often hereditary. Still, no one can predict who will become addicted to a particular drug. As simplistic as it may sound, then, "just say no" might be the answer!

Society's Problem

In the United States, the problem of illegal drug use has only grown worse over the years. In 1962, four million Americans had tried an illegal drug. According to data from the 1999 Substance Abuse and Mental Health Administration survey, 88 million Americans age 12 or older (40 percent of the population!) reported illicit drug use at least once in their lifetime, 12 percent reported use of a drug within the past year, and 7 percent reported use of a drug within the past month. Illicit drug use by 13- and 14-year-olds has been increasing as well. The following table shows the reported drug and alcohol use for high school seniors in 2000.

Reported Drug and Alcohol Use by High School Seniors, 2000

Drugs	Used within the last 12 months*	Used within the last month*
Alcohol	73%	50%
Marijuana	37%	22%
Stimulants	11%	5%
Hallucinogens	8%	3%
Other opiates	7%	3%
Sedatives	6%	3%
Inhalants	6%	2%
Tranquilizers	6%	3%
Cocaine	5%	2%
Steroids	2%	1%
Heroin	2%	1%

Numbers have been rounded.

Source: University of Michigan, Drug Use from the Monitoring the Future National Results on Adolescent Drug Use: Overview of Key Findings 2000, 2001.

In 1999, a total of 11,651 deaths related to drug abuse were reported in just 40 metropolitan areas. Cocaine is the most common drug reported in emergency department visits (nearly 170,000 in 1999) followed by accidents related to marijuana use with nearly 90,000 visits and heroin with more than 80,000 visits.

The drug epidemic also affects the workplace because drug users take off more time and file more claims for benefits and workers' compensation. People on drugs are also more likely to cause accidents, engage in domestic violence and child abuse, and commit other violent crimes. All of this costs the nation billions of dollars in health costs. The Office of National Drug Control Policy estimates that illegal drug use costs the nation $110 billion annually, but this number does not begin to approximate the emotional cost of drug abuse.

IQ Points

Some scientists believe that all addictions have a biological basis and therefore could be treated with neurotransmitter "cocktails." The research has not yet reached the point where this kind of treatment is possible, so treatments therefore remain largely confined to behavioral therapy.

Code Blue

Warning signs that young people may have problems with drugs are changes in school performance, such as falling grades, behavior issues, and truancy; increased family conflict; poor hygiene and changes in dress; and chronic health issues, such as anxiety, depression, and fatigue. Adults may have similar problems with health and family as well as difficulties with coworkers and employers.

Alcoholism

Anyone who abuses alcohol may develop a serious problem and ultimately suffer from alcoholism. Approximately 14 million people in the United States (1 in every 13 adults) suffer from alcohol abuse or dependence, with the highest rates found among people between the ages of 18 and 29 and the lowest among seniors 65 and older. Alcoholism definitely has a genetic component; for example, children of alcoholics are four times more likely than other people to become addicted.

The amount of alcohol that reaches the brain from a drink depends on a number of factors, including type and amount of alcohol in the drink, the size and weight of the person drinking, and whether the person has eaten recently. Unlike many drugs, alcohol is widely distributed in the brain, but it does target specific regions, notably the cortex, brain stem, and cerebellum. As you know from earlier in the book, these areas are involved in decision-making, balance, memory, and emotion.

Interestingly, a low dose of alcohol stimulates brain activity. Alcohol specifically increases the activity of GABA, an inhibiting neurotransmitter that affects the release of dopamine. Alcohol prevents the inhibition effect of GABA and therefore increases the production of dopamine. By inhibiting the effect of GABA, alcohol causes more dopamine, which is one of the neurotransmitters associated with pleasurable feelings, to be released. Thus, a couple of drinks produces a high. The initial impact is on the part of the brain that normally exercises self-restraint. The drinker feels more relaxed and less inhibited and his or her thinking may become impaired.

As more alcohol is consumed, however, the brain becomes less active, and the regions controlling motor and cognitive functions become impaired. The slightly intoxicated person now may become moody, have slurred speech, experience a loss of coordination, and have blurred or double vision.

If still more alcohol is ingested, brain activity declines even more. Intoxicated people may be unable to stand, walk, or control their bodily functions. They may vomit, become depressed, and pass out. In the most extreme cases, they may stop breathing.

One interesting aspect of alcohol's effect on the brain is that it can cause almost diametrically opposed reactions during one bout of drinking, that is, a high and a tranquilizing effect. One possible explanation for this is that alcohol inhibits another neurotransmitter, NMDA, which is involved in exciting neuron activity. Why aren't people high and sedate simultaneously? The answer is likely the varying sensitivity of alcohol in different parts of the brain.

Brain damage may occur from long-term alcohol abuse. Studies have shown that the gray matter of the brain shrinks in alcoholics and that alcoholics may lose short-term memory and have difficulty learning new information. If the person stops drinking, these functions may improve.

Alcoholics may also suffer metabolic diseases. Wernicke's encephalopathy, for example, is a condition caused by a thiamine (vitamin B1) deficiency. It is seen in chronic alcoholics and individuals suffering from malnutrition. It is characterized by confusion, unsteady gait (ataxia), and paralysis of eye muscles due to damage of the nerves going to these muscles (ophthalmoplegia). The symptoms clear up after the patient is treated with vitamin B1. The eye problems usually improve within 24 hours. The confusion and gait disturbance may take several days or even weeks to get better. Sometimes, the person never fully recovers their normal manner of walking.

Another condition that may be seen following Wernicke's encephalopathy is Korsakoff's syndrome. This is also seen in cases of chronic alcoholism and severe malnutrition. A vitamin B1 deficiency is again the cause, with the result being a marked decrease in short-term memory and, to a lesser extent, long-term memory. In most cases, large doses of thiamine can reverse the symptoms.

Gray Matter

No level of alcohol consumption is safe during pregnancy, but pregnant women who abuse alcohol may have babies who suffer from fetal alcohol syndrome. This syndrome includes birth defects such as distorted facial features, an abnormally small head and body, mental retardation, and poor hand and eye coordination. In addition, moderate drinking can decrease fertility

Nicotine

On April 14, 1994, the chairmen of the leading tobacco companies testified before Congress that nicotine is not addictive. Documents were unearthed by anti-smoking groups and lawyers involved in lawsuits against the industry, however, that indicated the industry's scientists were aware of nicotine's addictive properties as early as 1963, a year before the U.S. Surgeon General issued the first government report linking smoking and lung cancer. Had the Surgeon General known about nicotine and addiction before publishing the report, it might have helped change the behavior of many people and saved many lives.

When used for the first time, nicotine can cause nausea, but most users get used to the feeling and experience mild stimulation. Nicotine may enter the bloodstream through chewing tobacco, inhaling snuff, or smoking cigarettes. When smoked, nicotine reaches the brain almost immediately. For this reason, nicotine skin patches are sometimes used to reduce dependence on the drug because it takes longer for the nicotine to reach the brain and levels fall more gradually.

Nicotine is similar in structure to the neurotransmitter acetylcholine and affects the cortex, thalamus, cerebellum, and regions of the brain that control muscle contractions, thinking, and, in some cases, emotions. Nicotine primarily damages areas outside the brain, but it can increase the use of sugar in certain parts of the brain and affect acetylcholine receptors.

The addictive properties of nicotine make it difficult for people to quit smoking. The potential health consequences of smoking, such as lip, mouth, throat, lung, breast, and other forms of cancer, as well as heart disease and strokes are caused by tars and other chemicals in the cigarettes rather than the nicotine. Cigarette smoking is the chief preventable cause of death.

IQ Points
About 26 percent of the U.S. population smokes cigarettes, which is approximately 24 million men and 22 million women.

Marijuana

Marijuana is the most commonly used illegal drug. It is produced by drying the leaves and flowering tops of the cannabis plant (from which marijuana takes its name). Marijuana is typically smoked in a loosely rolled cigarette known as a joint.

The active ingredient of marijuana is THC (tetrahydrocannabinol), which resembles the natural brain substance anandamide. Sinsemilla, which comes from the unpollinated female cannabis plant, and hashish, made by taking the resin from the leaves and flowers of the marijuana plant and pressing it into cakes or slabs, have higher concentrations of THC than ordinary marijuana. THC lodges in the regions with anandamide, notably the cortex, cerebellum, amygdala, and hippocampus.

Marijuana can produce a dreamlike state, distort perceptions of time and space, increase appetite, and decrease concentration and coordination. The long-term effects of marijuana use remain unknown. Chronic use has been shown to cause apathy, dullness, and a loss of memory and concentration. Research on animals has suggested that marijuana abuse can lead to a decline in nerve cells and damage to connections in areas of the brain related to memory, such as the hippocampus. However, there is no conclusive evidence of permanent brain damage in humans as a result of marijuana use. Marijuana does cause damage to other parts of the body, however, such as the lungs. Recent research also suggests that the body may build up a tolerance to THC, which would make the medicinal use of marijuana less effective than other medications.

More and more teens are using marijuana these days. This increased use of marijuana is probably related to the growing sense among teenagers that it is not dangerous. According to a Partnership for a Drug-Free America survey, 72 percent of teenagers in 1990 thought marijuana was harmful. A decade later the figure had fallen to 54 percent. This statistic is particularly alarming in light of the fact that the potency of marijuana is increasing. According to the Drug Enforcement Administration, the average THC content of marijuana in the 1970s was 1.5 percent, but it now averages 7.6 percent.

Cocaine

Cocaine is the most powerful stimulant drug that occurs naturally; it is extracted from the leaves of the coca plant and is usually distributed as a white crystalline powder. Columbia produces 75 percent of the world's cocaine, and Americans buy it. About 3.7 million Americans (1.7 percent of people over 12) use cocaine, and about 1.5 million do so regularly.

Cocaine can produce feelings of energy, pleasure, and power. Users under the influence of cocaine typically experience a loss of appetite and difficulty sleeping. After the drug wears off, the user may be tired and depressed, and may crave more of the drug.

The speed and intensity of the effect of cocaine is partially determined by how it enters the body. Snorting a powdered form of cocaine results in the drug being absorbed through the mucous membranes of the nose and traveling through the circulatory system to the heart and ultimately to the brain. *Crack* cocaine enters the brain more quickly and at higher levels. The most direct route to the brain, which also produces the highest levels of drug in the bloodstream, is intravenous injection of the drug.

Words of Wisdom

Crack cocaine has been processed in a ready-to-use form for smoking. The term refers to the crackling sound heard when the mixture is heated.

Cocaine spreads throughout the brain, but it concentrates in regions that produce dopamine, the neurotransmitter associated with feelings of pleasure. Concentrations are also high in areas of the brain that control motion and affect mood. Meanwhile, the parts of the brain responsible for higher functions (thinking, learning, memory) receive less blood. Cocaine produces intense electrical activity in the brain, but when the effect wears off, the level of activity is abnormally low, which may explain why users become depressed after the high passes. Researchers also believe that cocaine may lead to a reduction in the number of dopamine receptors in the brain and that this reduction may also contribute to the continued craving that addicts feel when they try to stop using the drug, which may help explain the tendency for people trying to kick the cocaine habit to suffer relapses.

Chronic use of cocaine can reduce the brain's ability to use glucose, a vital chemical needed for energy, and, therefore, may reduce the brain energy level in people who abuse the drug. In addition, the drug stimulates chemicals in the brain that produce unhappy feelings, which can create a hunger for the drug to bring back the pleasurable sensation.

The most serious consequence of cocaine abuse is damage to the heart, but cocaine abuse can also cause seizures, strokes, and brain hemorrhage. Users who smoke or inject cocaine may be at even greater risk than those who snort it.

Percent of High School Seniors Reporting They Could Obtain Drugs Fairly Easily or Very Easily, 2000

Drug	Percentage*
Marijuana	89%
Amphetamines	57%
Cocaine	48%
LSD	47%
Crack	43%

continues

Percent of High School Seniors Reporting They Could Obtain Drugs Fairly Easily or Very Easily, 2000 (continued)

Drug	Percentage*
Barbiturate	37%
Tranquilizers	34%
Heroin	34%
PCP	29%
Crystal metham-phetamine (ice)	28%

Numbers have been rounded.

Source: University of Michigan, Drug Use from the Monitoring the Future National Results on Adolescent Drug Use: Overview of Key Findings 2000, 2001.

Heroin

Opiates, of which heroin is one, are effective pain killers. They can mimic naturally occurring body chemicals that block pain and produce pleasure. Morphine is a legal drug used for the treatment of pain. The illegal form of morphine is heroin, which changes to morphine once it's inside the body. Heroin is a derivative of the opium poppy plant. Pure heroin is a white powder with a bitter taste.

Like most drugs, heroin can be taken in a variety of ways that affect how quickly it reaches the brain. If heroin is taken orally, for example, it takes approximately 30 minutes to reach the brain and take effect. Snorting a powdered version causes the drug to be absorbed into the blood and reach the brain more quickly, so the effect is felt in about 10 to 15 minutes. Heroin is probably most commonly injected into a vein or muscle, and this method takes only about 15 to 30 seconds to reach the brain (which is why doctors give drugs intravenously). Partly because of AIDS, many heroin users fear needles and smoke a form of heroin that can transit through the lungs to the brain in only about 7 seconds and has both a high risk of overdose and addiction.

Opiates such as heroin concentrate in certain parts of the brain, namely the medulla, the floor of the fourth ventricle (the Area Postrema), and the brain tissue around the passageway from the third to the fourth ventricle where the body's natural opiates (enkephalins and endorphins) are found. These chemicals are involved in regulating the signals between neurons to decrease perceptions of pain and produce pleasurable sensations.

Heroin can cause euphoria, drowsiness, constricted pupils, nausea, and respiratory depression. There is also a significant risk of major stroke, caused by blood clots, which come from infection within the heart or constricted or inflamed blood vessels within the brain. Heroin can cause the clots, and cocaine can cause the narrowed and inflamed blood vessels.

Chronic use of opiates can lead to addiction, and even one-time use can be life-threatening. Overdoses of heroin can cause slow and shallow breathing, convulsions, coma, and death. Heroin overdoses have taken the lives of many famous people, including comedian John Belushi (who also was on cocaine), Jim Morrison of the Doors, and Janis Joplin. Heroin addicts are not just in danger from overdosing, however. They also are at high risk for AIDS and other viral and bacterial infections and often must resort to crime to get the money they need to maintain their habit.

Short-term, regular morphine or heroin use has a mild effect on the brain's ability to use glucose. However, longer-term exposure to morphine doesn't affect the use of glucose, which means the brain adapts to the constant presence of morphine. If an addict stops taking morphine, the brain doesn't return to normal; instead, the brain's use of glucose increases to extreme levels. To maintain normal brain function, some addicts use opiates such as methadone. Methadone is a synthetic drug that does not produce the same high as heroin, but it prevents withdrawal and the craving for heroin. Methadone is used in clinics to help known addicts escape the dependence on illegal drugs and adopt a drug-free life.

Code Blue

Popular movies have created an inaccurate image of Vietnam veterans as more likely to be drug addicts than others in their age group. As many as 20 percent of soldiers serving in Vietnam may have used heroin because it was easily accessible. Once the soldiers returned home, however, the drug was not readily available, so only a small number, about 1 percent, remained addicted.

Code Blue

Cocaine and heroin make a deadly combination. A "speed ball" is a combination of heroin and cocaine, which can produce a high that has far greater intensity than either drug by itself.

As is the case with marijuana, another disturbing trend related to heroin is the increased purity of the drug today. Heroin used to have purity levels ranging between 1 and 10 percent, but the national average purity level today is 35 percent. The DEA says South American heroin can be as much as 80 percent pure.

Hallucinogens

Although the government and educational system have had limited success convincing a large segment of the population that other illicit drugs are dangerous, greater progress has been made in discouraging the use of *hallucinogens*. Hallucinogens (also known as psychedelics) are drugs that affect a person's perceptions, sensations, thinking, self-awareness, and emotions. These drugs include LSD, mescaline, PCP (phencyclidine or "angel dust"), and psilocybin.

IQ Points

The use of hallucinogens as a class has fallen since the 1980s largely because people believe there is a great risk from taking drugs like LSD even once and that it is especially harmful if taken regularly.

Some hallucinogens come from natural sources, such as mescaline from the peyote cactus and psilocybin from certain types of mushrooms. Others, such as LSD (lysergic acid diethylamide), are synthetic or manufactured.

Hallucinogens affect regions of the brain that produce serotonin. In particular, they affect the cortex, thalamus, and reticular activating system. These parts of the brain control sensory perception, thinking, movement, sexual behavior, and sleep.

Heavy users of hallucinogens sometimes develop signs of brain damage, such as impaired memory and concentration, mental confusion, and difficulty with abstract thinking. It is not yet clear whether discontinuing the use of LSD and other hallucinogens can restore these functions or whether the damage is permanent.

LSD

LSD was discovered in 1938 and is the most potent hallucinogenic. During the 1950s and 1960s, scientists used LSD on patients to study mental illness because of its structural resemblance to a chemical in the brain and the similarity of its effects to certain aspects of psychosis.

LSD is odorless, colorless, and tasteless and is referred to on the street by more than 80 different names. The drug can be found in a variety of forms, including on sugar cubes, in squares of gelatin, and on sheets of paper soaked with the chemical.

The effect of LSD and other hallucinogens is unpredictable and can be affected by the amount taken, the user's mood, and the surroundings in which the drug is used.

Usually, the user feels the first effects of the drug 30 to 90 minutes after taking it. The physical effects include dilated pupils, higher body temperature, increased heart rate and blood pressure, sweating, loss of appetite, sleeplessness, dry mouth, and tremors. Sensations and feelings change, too. Users often report a distorted perception of time and space and may feel several different emotions at once or swing rapidly from one emotion to another. Users say they can "hear" colors and "see" sounds. Judgment is impaired. These experiences are referred to as a *trip*.

After an LSD trip, users may feel anxious, scared, and depressed. Days or even months after taking the drug, a user may experience a *flashback*, which is a recurrence of the drug's effect.

PCP

PCP was first developed as an anesthetic in the 1950s, but it was taken off the market for human use because it sometimes caused hallucinations. PCP is produced in either white, crystal-like powder or a tablet or capsule. It can be swallowed, smoked, sniffed, or injected. Effects include increased heart rate and blood pressure, dizziness, sweating, and numbness. Large doses can cause drowsiness, convulsions, coma, and death.

PCP can cause users to become violent and behave in bizarre and dangerous ways. The drug can have the opposite effect on different people, causing one to become withdrawn and uncommunicative and another to be aggressive. People under the influence have died in falls, automobile accidents, drowning, and a variety of other ways resulting from impaired judgment. Regular PCP use affects memory, perception, concentration, and decision-making. Users may show signs of paranoia, fearfulness, and anxiety. Some effects can linger for days or weeks. Long-term use of PCP can cause memory and speech difficulties as well as delusions, such as hearing voices.

MDMA

Another drug with hallucinogenic properties is MDMA (which has a really long chemical name) or Ecstasy. This drug was first made in 1912 by a German drug company, apparently as an appetite suppressant. It has become popular as a recreational drug only in the last 20 years or so and is now popular in nightclubs, rock concerts, and late-night parties called *raves*. Though Ecstasy abuse is not as widespread as that of other drugs, it is increasing significantly.

Ecstasy usually comes in a tablet or capsule. The drug stimulates the release of serotonin in the brain and produces extreme relaxation, empathy for others, and positive feelings. It can also suppress basic needs for food and sleep. The effect lasts for four to six hours. Ecstasy is not as addictive as heroin or cocaine, but it is associated with numerous harmful effects, including hallucinations, chills, sweating, nausea, tremors, blurred vision, involuntary teeth clenching, and increases in body temperature. Overdoses can cause seizures, high blood pressure, faintness, panic attacks, loss of consciousness, and death. Research into the long-term effects of Ecstasy suggest that even recreational users may be at risk for developing permanent brain damage and may suffer from depression, anxiety, memory loss, and other neuropsychotic disorders.

Uppers and Downers

Amphetamine, speed, meth, and diet pills (such as dexedrine and benzedrine) are powerful stimulants that cause the release of dopamine. This release floods the receptors and causes sensations of extreme energy, pleasure, and alertness, which is why users typically take them to stay awake, to maintain high levels of energy, and to increase the ability to work. Used properly, under a physician's care, amphetamines can be beneficial, and they are regularly prescribed for a variety of problems, including narcolepsy (a condition marked by the sudden and urgent need to sleep).

Chemically, stimulants look similar to dopamine and concentrate in the areas of the brain where that neurotransmitter is produced, such as the basal ganglia, cortex, and hippocampus. They increase heart and respiration rates, increase blood pressure, dilate the pupils, and decrease appetite. Side effects can include anxiety, sleeplessness, dizziness, and blurred vision. They can also produce delusions, hallucinations, paranoia, and bizarre behavior. Stimulants can be addictive as users seek to avoid the depressed feeling that follows when the drug wears off by taking more of the drug or even stronger stimulants.

The long-term impact of stimulants is that the nerve endings with serotonin may be damaged or lost, but there is no permanent impact on brain function. Negative effects usually dissipate when drug use is discontinued.

Tranquilizers, such as diazepam (Valium), lodge in GABA receptors and widen the ion channels on cell membranes of neurons. This change may impair coordination, induce sleepiness, reduce anxiety, relax muscles, and lead to depression and a loss of consciousness. The effect is to dampen brain activity, particularly in the cortex, midbrain, and cerebellum.

Code Blue _____

Do you have a drug or alcohol problem?

If you answer "yes" to one or more of these questions (taken from *Changing Your Mind: Drugs in the Brain* by Bertha K. Madras, Ph.D.), then you may have a problem with drug abuse or addiction:

1. Do you feel the need to cut down on drug use?
2. Are you annoyed when people comment on your drug use?
3. Do you use drugs early in the day?
4. Do you use drugs alone?
5. Does anyone think you have a problem with drugs?
6. Have you gotten into trouble from using drugs?
7. Do you use drugs to feel better about yourself?
8. Do you ever regret things you did while using drugs?
9. Do you ever use alcohol or drugs to change your mood?
10. Do you ever use drugs to fit in with friends?

The "Date Rape Drug"

Drugs have long been used in an effort to enhance sexual performance, intensify feelings of pleasure, and lower the inhibitions of the opposite sex. Over the years, men in particular have given women drugs such as Quaaludes (sometimes overtly and sometimes surreptitiously) to put them in a state that affects their judgment and makes them more willing to have sex or renders them essentially helpless to prevent someone from taking advantage of them.

Today, a new drug is becoming widely abused and has become associated with date rape. The drug is flunitrazepam, better known by its brand name Rohypnol or its street moniker, "Roofies." Rohypnol is similar to Valium, but it is a far more powerful tranquilizer that can cause amnesia, muscle relaxation, dizziness, nausea, and disorientation. The drug usually knocks out the user in 20 to 30 minutes and lasts for several hours.

Rohypnol may be legally prescribed for severe insomnia and to sedate people who suffer from psychoses or are about to undergo surgery. It is being illegally used by people who want to incapacitate and sexually assault women who then are unable to remember what happened to them. The drug is being used, for example, to spike drinks at parties and social gatherings. Besides the physical and psychological dangers related to having sex while under the influence of Rohypnol, the drug can also be addicting if used repeatedly and, when combined with alcohol or other drugs, may cause severe physical damage and can result in death.

The Least You Need to Know

- Illegal drugs can be dangerous because they mimic the activities of neurotransmitters and alter the normal communication in the brain.

- How many people use a particular illegal drug is related to the perception of the drug's long-term harm. Use of LSD has declined because people believe taking it even once is dangerous while marijuana use has increased because it is not viewed as harmful.

- Abuse of alcohol can be a serious health hazard that affects not only the drinker, but those around him or her. Children of alcoholics are at a much greater risk of becoming alcoholics themselves.

- Predicting whether someone will become an addict is impossible, but drugs such as heroin and cocaine stimulate cravings for the good feelings they produce that can lead even a one-time user to get hooked.

20

Crossed Wiring

In This Chapter

◆ Knowing the difference between disorders and diseases
◆ Breaking with reality
◆ Experiencing big-time anxiety
◆ Feeling down—and up
◆ Getting your kid's attention

According to the Surgeon General of the United States, about one in every five people suffers from a mental illness in a given year, but two thirds of them do not seek treatment. Mental illness can affect people of all ages. Nearly 21 percent of children between 9 and 17 have some mental illness, as do nearly 20 percent of those 55 and over. That's a lot of people—and a scary thought, considering how mental illness affects your functioning.

Mental illness may cause abnormalities in emotions, thinking, and perception, and the ability to interact with others. It can range in severity from moderate, as in the case of mild depression, to disabling, as is the case with psychoses. People suffering from mental disorders may be unable to live normal lives and in the worst cases may require hospitalization because they can't take care of themselves.

This chapter provides an overview of the different types of mental illness.

It's All in Your Head

To diagnose a mental illness, physicians must take a careful medical history and perform a neurological exam, but they usually do not require more extensive medical testing. However, many mental illnesses (also known as mental *disorders*) have symptoms that are similar to or the same as organic brain *disease*, such as the hallucinations that may be seen with the late stages of many infectious or degenerative diseases of the brain. Because the symptoms of mental illness often mimic those of organic brain disease (for example, brain tumors), in some cases it may be necessary to use diagnostic studies such as MRI scans to rule out a physical cause of the problem. Having a mental illness can be traumatic, but having it misdiagnosed can throw you over the edge.

Current research suggests that the symptoms of mental illness may be related to pathological changes in the brain and, more specifically, various chemical imbalances that may exist. Much research focuses also on the amygdala and hippocampus, which influence emotion and the fight-or-flight response.

Words of Wisdom

Mental health conditions are usually called disorders rather than diseases. A **disease** is a condition with detectable physical changes whereas a **disorder** typically has unknown pathology and is associated with an impairment of functioning.

Here's a typical case a neurosurgeon might encounter: John visits the doctor because his wife says his personality has changed and he acts differently than he used to. He can't remember where he puts his keys when he never had trouble finding them before. He always treated his mother-in-law nicely, but he now is rude and mean to her. If John is elderly, the doctor's first suspicion might be Alzheimer's, but if he is 45 or younger, that diagnosis is far less likely.

The doctor performs a neurological exam and finds no evidence of paralysis or cranial nerve damage, so the only problem appears to be the personality change. The doctor knows that personality changes can occur from brain tumors affecting the temporal lobes, limbic system, hippocampus, and other structures involved in memory and personality and may decide to do an MRI scan to rule out a tumor, multiple sclerosis, or a degenerative disease.

During the discussion with the patient and family, the doctor may learn that John's family has a history of mental illness, perhaps he had a brother who committed suicide or was schizophrenic. If such a family history exists, the likelihood that John suffers from a mental disorder is high. His doctor prescribes an MRI, which is normal, and so the doctor diagnoses Alzheimer's disease.

From Behind Closed Doors

For centuries, people believed that humans' mental condition was in the hands of the gods and that people with mental illness were possessed by the devil. This misperception led to frequent persecution of those who acted "disturbed," and for some time the church hierarchy tied this persecution to fears about witchcraft and sorcery. Yet even during periods when these beliefs were dominant, some physicians suggested that mental illness could be traced to natural rather than supernatural causes.

Try a Little Tenderness

The mentally ill were not viewed as people who needed help, but as people who needed to be freed of the evil that possessed them, even if that killed them. *Lunatics* were locked away in asylums, chained, and beaten.

Bethlehem Hospital in London, founded in 1247, was one of the first hospitals for the mentally ill (they weren't hospitalized there until 1377). The hospital became known as Bedlam, and people treated it almost like a tourist attraction, coming for visits to observe the behavior of the people confined there. (This is where the word *bedlam* comes from.) Today there are so many mentally ill people roaming the streets that you don't have to go to a hospital to see such bedlam.

Words of Wisdom

The word **lunatic** is derived from the idea that a person could become deranged during certain lunar phases from the influence of the moon.

The first mental hospital in North America opened in 1752 in Philadelphia. The patients were chained to the walls of prison-like cells, and straitjackets were introduced to prevent patients from injuring themselves or others.

Not until the nineteenth century did the idea of treating the mentally ill humanely and trying to help them get better become accepted. Emil Kraepelin, a German psychiatrist, published the first classification system for mental disorders in 1883. He and other scientists began to understand that unusual behavior might have organic or psychological explanations based on the working of the mind rather than simply being the work of demons. Nevertheless, to this day, a stigma is attached to people who have mental illnesses and seek out therapists or psychiatrists because many people view these individuals as people who can't solve their own problems, cope with life, or are in some other way lacking the fortitude to live normal lives rather than seeing them as people with illnesses that require treatment.

Into the Twentieth and Twenty-First Centuries

After World War II, an increasing number of mental health professionals came to the conclusion that only the most severely disturbed and disabled people should be institutionalized. At about the same time, new drugs were introduced that made it possible to control many of the symptoms and allow people suffering from even severe disorders such as schizophrenia to live outside of hospitals and, in many cases, lead relatively normal lives.

> **Words of Wisdom** _____
>
> Many people make pilgrimages to the holy city of Jerusalem. Some of these people become so intoxicated by the spirituality of the place that they begin to act erratically, and some even develop temporary or long-term delusions of grandeur, such as the belief that they are the messiah or that they can speak directly to God or Jesus. This occurs so often that psychiatrists recognized it as an illness they call the **Jerusalem syndrome.**

The classic understanding of mental illness in modern times is that it has no underlying physical cause. At the other extreme is the newer school of thought that mental illness is just a physical disease with mental symptoms. Perhaps the best expression of the current thinking, which combines the two ideas, comes from the surgeon general's suggestion that a person's mental state is the product of that person's life experiences and genetic inheritance.

No Shortage of Disorders

A large number of mental illnesses exist, but this chapter focuses primarily on the most common and debilitating ones. Each mental condition that is discussed has a list of symptoms, but you should not jump to the conclusion that you are ill if you suffer from any of the listed symptoms. Mental illness is usually characterized by a combination of multiple symptoms. If you do recognize several on one or more of the lists in your own feelings and behavior, you may want to consult a doctor.

All mental disorders are grouped into these major classifications:

- Disorders usually first diagnosed in infancy, childhood, or adolescence (mental retardation, ADHD)

- Delirium, dementia, amnestic (loss of memory), and other cognitive disorders that involve a significant deterioration of mental functioning (Alzheimer's disease)

- Mental disorders due to a general medical condition (personality change not related to a specific disorder)

- Substance-related disorders (related to drug abuse)
- Psychotic disorders (schizophrenia)
- Mood disorders (depression, bipolar disorder)
- Anxiety disorders (panic, phobias, obsessive-compulsive disorder, post-traumatic stress disorder)
- Somatoform disorders involve physical symptoms arising from psychological problems (a person says that he is blind or deaf, but nothing is wrong with his eyes or ears)
- Factitious disorders (people who do things to themselves or fake symptoms to appear sick)
- Dissociative disorders that involve distur-bances in a person's consciousness (amnesia)
- Sexual and gender identity disorders (fetishism, exhibitionism, confusion over sexual orientation)
- Eating disorders (anorexia, bulimia)
- Sleep disorders (insomnia)
- Impulse-control disorders (kleptomania, pyromania)
- Adjustment disorders (extreme emotional reaction to an event within the last three months)
- Personality disorders (maladaptive personal-ity traits)

> **Words of Wisdom**
>
> What ever happened to what was once thought of as sim-ply neurotic behavior? The word **neurosis** was applied to disorders that had no apparent organic cause. Basically all the mental ill-nesses were once referred to as neuroses, but the word has fallen out of favor, and mental health professionals rarely use it now. The word is mainly reserved to describe characters in Woody Allen movies.

Psychosis and Schizophrenia

Psychosis is the general category of mental illness that involves disturbances of perception and thought processes. The most common symptom is hallucination, which is the belief that a sensory experience is real when it is not. Hearing voices that are not there is probably the most typical hallucination, but hallucinations also can be visual (sight), olfactory (smell), tactile (touch), or gustatory (taste).

Another common symptom of psychosis is delu-sion, which is an erroneous belief that a person holds despite evidence to the contrary. One type of delusion is paranoia, in which patients are con-vinced that others are out to get them by poison-ing, cheating, or conspiring against them.

> **IQ Points**
>
> We can't help but be reminded of former secretary of state Henry Kissinger's famous line, "Even paranoids have enemies."

Sometimes delusional people think they are receiving messages from aliens or through radio waves. Another stereotypical symptom is the delusion of grandeur, in which people believe that they are some famous person such as Jesus or Napoleon or have qualities just like that famous person.

People do not suddenly develop a psychosis. Usually the illness develops slowly over time. The early warning signs often are not recognized until after the person begins to exhibit the more serious symptoms.

Schizophrenia

The most extreme psychosis is schizophrenia, a devastating disorder that typically appears during the teen years and early twenties, which is a key time in life when people are establishing social relationships and careers. More than two million Americans (about 1 out of every 100 people over 18) have schizophrenia. Unlike most mental illnesses, this rate of incidence is remarkably stable across cultures.

In addition to delusions and hallucinations, schizophrenia is characterized by disorganized, bizarre, and illogical thought processes that produce similarly disorganized, bizarre, and illogical behavior. In addition, patients often have difficulty expressing emotion and reacting appropriately and are fearful and withdrawn. One extreme symptom is to withdraw so completely that the person does not move and perhaps does not speak for extended periods. This behavior is what most people associate with catatonic schizophrenia, but catatonics may also display uncontrollable mannerisms and activity. Schizophrenics are also at a higher than average risk of committing suicide, with about 10 percent taking their own lives.

Some schizophrenics suffer symptoms sporadically and can lead rather normal lives in between; others have constant or recurring symptoms that are disabling. The following symptoms are common among schizophrenics:

- ◆ Hallucinations
- ◆ Delusions
- ◆ Agitation
- ◆ Disorganized thoughts and behaviors
- ◆ Flat or blunted affect (emotion)
- ◆ Difficulty with abstract thought
- ◆ Inability to experience pleasure
- ◆ Illogical thoughts
- ◆ Poor motivation, spontaneity, and initiative

Bad Genes?

Research indicates that schizophrenia has a genetic component and may be related to a chemical imbalance involving the neurotransmitters dopamine and glutamate. A single, "bad" gene may not cause the illness, but rather one or more genes may make some people more susceptible to the illness. A person with a parent or sibling with schizophrenia has about a 10 percent risk of developing the illness; a person with no family history of schizophrenia has a 1 percent risk for developing the disease.

The cause of schizophrenia is not entirely genetic, however, because studies of identical twins have found that there is only a 50 percent chance that both will have schizophrenia. This finding has led scientists to theorize that environmental factors, especially during fetal development, may play a key role.

Thanks to modern imaging techniques, scientists have also found evidence that some people with schizophrenia have enlarged ventricles in their brains. There also appears to be a relationship between schizophrenia and especially high levels of dopamine.

> **IQ Points**
>
> The movie *A Beautiful Mind* does a very good job giving a sense of what it must be like to be schizophrenic; however, one misleading aspect of the story about the Nobel Prize-winning economist John Nash Jr. was the implication that a person can overcome the disease by pure force of will. A schizophrenic rarely functions in society without the help of medication.

Beautiful Minds

For many years little medical treatment was available to ameliorate the most severe symptoms of schizophrenia. Antipsychotic drugs such as haloperidol (Haldol) and chlorpromazine (Thorazine) were found to moderate hallucinations, delusions, and the reduced motivation and blunted expression many schizophrenics suffer, but these drugs often had severe side effects, in particular, a disorder known as tardive dyskinesia (TD), which is characterized by involuntary movements. Newer drugs called atypical antipsychotics, such as risperidone (Risperdal) and olanzapine (Zyprexa), have proven even more effective in treating the symptoms of schizophrenia with fewer side effects. Unlike the stereotypical mental patients medicated on television and movies, schizophrenics taking these drugs do not become barely conscious vegetables. Yet the drugs do not cure the illness or guarantee the schizophrenic will be free of psychoses either.

> **CAUTION**
>
> **Code Blue**
>
> Contrary to the popular image, schizophrenics do not have split or multiple personalities. Also, they are usually not violent toward others, though their often bizarre behavior may be frightening.

Medication combined with intensive therapy allows many schizophrenics to manage relatively normal lives. As with many other illnesses, early detection can help improve the chance of a better outcome. Unfortunately, the illness is such that schizophrenics are often unable to recognize that they are sick and need help. Even those who do follow a treatment program can easily relapse if they consciously decide to stop taking their medication or forget to take it. Clinicians have found that only about one in five schizophrenics recovers completely, but most continue to have difficulties with social interaction.

Anxiety

The line between what is considered normal mental health and mental illness is sometimes a fine one. Anxiety, for instance, is an important and normal emotional response to fear and stress. People begin to cross the line, however, when they begin to feel anxious for no particular reason and their thoughts become consumed with fear. This kind of response falls under the heading of an anxiety disorder, which afflicts an estimated 19 million Americans each year, making it the most common of all the mental disorders.

The following anxiety disorders are among the most common:

- Social anxiety disorder (SAD)
- Post-traumatic stress disorder (PTSD)
- Generalized anxiety disorder (GAD)
- Panic disorder
- Obsessive-compulsive disorder (OCD)

People suffering from social anxiety disorder (SAD) have extreme reactions in social settings. They may sweat, shake, or have difficulty breathing. Their fear of embarrassment or humiliation can be so strong that they have difficulty being around other people. More than 10 million people suffer from SAD. This condition should not be confused with another emotional disorder that is known by the same acronym, Seasonal Affective Disorder. That SAD is a form of clinical depression that affects about 10 million people who become so upset by the change of weather (usually as fall ends and it gets colder) and gloominess of the winter that they become dysfunctional.

Post-traumatic stress disorder (PTSD) was discussed in Chapter 15. This disorder is seen in people who have experienced a traumatic event that they cannot put behind them. They frequently have flashbacks and often a disabling level of anxiety when they are put in a situation that reminds them of the traumatic incident. More than 13 million people are thought to have PTSD.

Generalized anxiety disorder (GAD) is a persistent and excessive concern that is unjustified. Unlike a phobia, which is a fear specifically associated with a particular stimulus, GAD is anxiety about a broader range of experiences. A person with GAD always expects the worst, whether the issue is health, finances, or work. GAD afflicts about four million people, affects more women than men, and usually begins during childhood or adolescence. As with other anxiety disorders, GAD is usually accompanied by physical symptoms such as trembling, nausea, and headaches.

Once again, a chemical imbalance involving one or more neurotransmitters, in particular serotonin, may play a role in the illness. Some research also hints at a genetic component, but the evidence remains inconclusive.

GAD is usually treated with a combination of cognitive-behavioral therapy and medication. Cognitive-behavioral therapy teaches people how to reduce their anxiety.

The following signs signal acute anxiety:

- Feelings of fear or dread
- Rapid heart rate
- Lightheadedness or dizziness
- Trembling, restlessness, and muscle tension
- Perspiration
- Shortness of breath
- Cold hands and/or feet

> **Gray Matter**
>
> Insanity is not a medical description. People are mentally ill or healthy, and those with illnesses range in severity. Insanity is a legal term that means a person can't control his or her behavior and is unaware that his or her actions are wrong.

It's a Panic

As many as seven million people, a majority of whom are women, experience panic attacks. These brief, intense feelings of terror come on suddenly and cause stress-related physical reactions such as sweating, shaking, and accelerated heartbeat.

Research suggests that abnormal activity in the amygdala may contribute to panic disorders. Medication and cognitive-behavioral therapy are the typical treatments for panic disorder.

Can't Help Myself

People with obsessive-compulsive disorder (OCD) can't stop thinking or doing the same thing over and over. They have rituals they feel compelled to repeat. Approximately five million people have OCD. Unlike many of the other mental illnesses, OCD affects men and women equally.

In the film *As Good As It Gets*, Jack Nicholson played a wonderful fictional character that displayed the symptoms of OCD. Nicholson would not step on cracks in the sidewalk, he would turn the lock on his door a certain number of times whenever he closed it, and he had a medicine cabinet full of carefully arranged soaps that he would use to repetitively wash his hands.

The general perception is that people learn their obsessive-compulsive behaviors. This isn't necessarily so. For example, if parents are especially concerned about their children's hygiene, it does not mean that the kids will grow up with a hand-washing obsession. The evidence suggests that people with OCD have different patterns of brain activity in an area known as the striatum.

A number of medications have been proven effective in helping people with OCD. A type of behavioral therapy known as "exposure and response prevention" is also useful. This therapy exposes people to the things that trigger their ritualistic behavior or obsessive thoughts and instructs them how to handle the anxiety and thereby avoid performing the compulsive rituals that normally accompany their unease.

Phobias

A specific, persistent fear that is extreme and irrational and compels a person to avoid the situation or thing that triggered the fear is referred to as a phobia. People can develop phobias of virtually anything, and the name for the fear simply has the suffix *phobia*. For example, here are a few of the more unusual phobias:

- **triskaidekaphobia**   Fear of the number 13
- **peladophobia**   Fear of bald people
- **pediophobia**   Fear of dolls
- **acarophobia**   Fear of itching
- **geliophobia**   fear of laughter
- **catoptrophobia**   Fear of mirrors
- **metrophobia**   Fear of poetry
- **kathisophobia**   Fear of sitting down
- **placophobia**   Fear of tombstones
- **cacophobia**   Fear of ugliness

And, of course, there is always phobophobia, which is the fear of phobias.

More than six million Americans have phobias, the more common ones being fear of heights (acrophobia), fear of confined spaces (claustrophobia), fear of open spaces (agoraphobia), and fear of foreigners (xenophobia). These fears usually begin during childhood or adolescence and, without treatment, persist into adulthood.

Medication may help ease the anxiety associated with a phobia, but the fear itself is best addressed through therapy. Phobias respond quite well to treatment.

Beyond the Blues

Everyone feels down, sad, guilty, tired, and irritable sometimes, but when those feelings cannot be shaken and persist for days or months, they move beyond the typical blues and become depression. More than 20 million people suffer from depression annually, and it is one of the leading causes of disability in the world and the principal cause of suicide.

The most severe forms of depression are characterized by a combination of symptoms that make it difficult for a person to function. Depressed people may not be able to eat, sleep, work, or experience pleasure. Generally, to be characterized as a major depressive disorder, the symptoms have to last for at least two weeks. A person can suffer from this type of major depression once in a lifetime, or depression may be a recurrent problem.

The most common symptoms of depression include the following:

- Persistent sadness or despair
- Insomnia
- Decreased appetite
- Difficulty concentrating, remembering, or making decisions
- Anhedonia (inability to experience pleasure)
- Restlessness and irritability
- Apathy, poor motivation, and social withdrawal
- Hopelessness and feelings of guilt or worthlessness
- Persistent physical symptoms, such as chronic pain, digestive disorders, and headaches that do not respond to treatment
- Thoughts of death or suicide

> **IQ Points**
>
> Many famous people have admitted or were believed to have suffered from depression or bipolar disorder in which their mood swings from the extreme high of manic behavior to an equally extreme low of depression. In some cases, their most productive periods corresponded with manic episodes. Here are a few recognizable names from a long list: Winston Churchill, Vincent van Gogh, Georgia O'Keefe, Ernest Hemingway, Patty Duke, and Charlie Parker.

Some people experience symptoms similar to depression, but they are not as severe or disabling. *Dysthymia* is a type of mental illness that allows people to carry on their lives, but they generally do not function as well as they normally would, and they frequently feel unhappy.

Equal Opportunity Lows

People of all ages may suffer from depression. Though it is often assumed that one naturally becomes depressed with age, this is not the case. People who see aging as a natural part of life and remain active generally feel content and happy in their golden years.

On the other end of the chronological spectrum, children are not necessarily happy and carefree. They sometimes develop depression, but it is often misinterpreted as a phase or behavior problem. The signs may be that the child is angry or irritable, claims to be sick, or acts unusually clingy to the parent. Because children may not be very good at communicating their feelings, a diagnosis may be difficult, but dramatic changes in a child's behavior and mood warrant a visit to the school counselor, pediatrician, a social worker, and/or a psychologist.

Women Get Down, But So Do Men

Women suffer from depression twice as frequently as men, and women between 35 and 45 are especially susceptible. Their moods can be affected by their menstrual cycle, the menopausal cycle, pregnancy, and childbirth. A particular type of depression known as postpartum depression can be caused by the physical and hormonal changes that accompany the birth of a baby, combined with the stress of caring for a newborn.

> **CAUTION**
> **Code Blue**
> More women attempt suicide than men, but men are four times more likely to succeed because they typically use more lethal methods. For example, men are more likely to shoot themselves, a method which has a better chance of leading to death, than overdose on pills, a method which is more typically used by women.

Men are by no means immune to depression, but they are less likely to admit they have a problem and seek help. The symptoms of depression in men are also often different. Instead of becoming withdrawn and feeling helpless and dispirited, men often display anger and irritability.

Depression can be triggered by a traumatic event, such as an accident, a serious illness, or the death of a loved one. Some evidence has been found to suggest that depression can be inherited. It is also believed to be related to chemical imbalances involving the neurotransmitters serotonin and norepinephrine. Those who believe in the chemical basis for depression see the illness as similar to other diseases that require regular doses of medication to control the symptoms.

Treatments, Worts and All

Antidepressants such as fluoxetine (Prozac) have proven very helpful for many people who suffer from depression. Sometimes a number of different medications need to be tried before an effective one for a particular individual is found, and it usually takes several weeks for the medication to take effect. Antidepressants, like all drugs, can cause side effects, including constipation, blurred vision, dizziness, sexual problems, headaches, nausea, and dry mouth. Consult your doctor if you have any questions or concerns about medication for depression.

In addition to drugs, or sometimes as an alternative to them, many people enter one of the many forms of therapy. There has also been a movement toward the use of herbs to treat depression, the most well-known being St. John's wort. This plant has been used for centuries as a folk remedy and has become the principal antidepressant used in Germany. Scientific research has not conclusively shown that St. John's wort is effective in the treatment of depression; however, the National Institutes of Health is conducting a three-year study to evaluate the herb. Recent reports have suggested it may not be as effective as previously thought in the treatment of depression.

Gray Matter

Some scientists believe that suicidal behavior, as well as depression, may be related to decreased levels of serotonin in the brain.

Bipolar Disorder

At the other extreme from depression is mania, an abnormally and persistently elevated mood or irritability. When people experience these highs in cycles with the lows of depression, they may suffer from manic-depressive illness, more commonly known today as bipolar disorder. When feeling manic, a person may be a bundle of energy, overdoing everything. This mania can affect an individual's decision-making and lead to a variety of behaviors that are monetarily or psychologically costly or sometimes just plain embarrassing. Depressive mood swings usually occur more often and last longer than manic ones. People with this disorder may also have psychotic symptoms. The disorder afflicts more than three million people, nearly 20 percent of whom commit suicide. It typically strikes people around age 30.

The following symptoms are associated with mania:

- ◆ Persistently elevated or euphoric mood
- ◆ Grandiose notions
- ◆ Unusual irritability
- ◆ Decreased sleep
- ◆ Increased sexual desire
- ◆ Racing thoughts
- ◆ Inappropriate social behavior
- ◆ Poor judgment and impaired impulse control
- ◆ Increased talking

In the case of bipolar disorder, the popular media has once again helped shape attitudes and perceptions with a series of episodes on the popular drama *ER* featuring Sally Field as the mother of one of the nurses. The erratic behavior of Field's character was attributed to bipolar disorder. One of the messages of that story line was that the failure to stay on the prescribed medication typically leads to manic-depressive episodes.

Many researchers blame chemical imbalances for bipolar disorder. A deficiency of natural chemicals such as dopamine and norepinephrine may trigger depression while an excess may cause mania. Other scientists suggest that abnormalities in the structure and/or function of certain brain circuits are responsible for bipolar disorder. In addition, strong evidence of a genetic component in bipolar disorder exists: two-thirds of the people with bipolar disorder have at least one close relative who is clinically depressed or bipolar.

As in the case of depression alone, bipolar disorder is typically treated with a combination of medication and therapy. Lithium has been used for many years to treat bipolar disorder, and newer mood-stabilizing medications, known as anti-convulsants, include valproate (Depakote) and carbamazepine (Tegretol).

Autism

People with autism usually cannot communicate with others and seem to live in their own world. They sometimes engage in repetitive behaviors and body movements such as rocking and may develop excessive attachments to people or objects. They may also be aggressive toward others and/or inflict injuries on themselves.

The symptoms of autism usually appear by the time a child is three and stay with them for the rest of their lives. About 1 to 2 of every 1,000 people are autistic, with males four times more likely than women to have the disorder.

Autism is not hereditary. The exact cause is unknown, but scientists have found differences in the electrochemical activity and structure of the brain in autistic patients. Medication, therapy, and special education can help alleviate some symptoms.

ADHD

Attention-Deficit Hyperactivity Disorder (ADHD) has become an increasingly common diagnosis for children who have difficulty concentrating, sitting still, and controlling their impulsive behaviors. As the disorder has gotten more publicity, more and more children have been diagnosed with ADHD. It is now thought to affect as many as two million children (5 percent), with boys being two to three times more likely to be affected.

Some children grow out of the disorder, but many others do not. Adults are increasingly coming to the realization that they may have suffered from ADHD since childhood and are now seeking medication and therapy to help them overcome problems that now affect their jobs and social interactions.

Gray Matter

A joint study by the National Institute on Drug Abuse and the National Institute of Mental Health found that boys with ADHD who were treated with stimulants were significantly less likely to abuse drugs and alcohol when they got older. On the other hand, people with ADHD, particularly if untreated, are at greater risk for substance abuse.

ADHD is the subject of intensive research today because of its prevalence among children. No one is sure of the cause, but brain imaging studies have shown that certain brain structures (prefrontal cortex, striatum, basal ganglia) are smaller than average in people with ADHD and that the areas controlling attention are less active. Some researchers have also suggested environmental causes, such as lead, may play a role in the disorder. Also, the finding that people with ADHD usually have a close relative with the same problem suggests the disorder is hereditary.

As is true for many other mental illnesses, medication and therapy can ameliorate the symptoms of the disorder. Today, millions of children diagnosed with ADHD are taking medications such as methylphenidate (Ritalin) and amphetamines (Dexedrine and Adderall). It may seem counterintuitive to prescribe stimulants to hyperactive children, and researchers don't have great explanations for why they work, but the basic theory is that the areas of the brain involved in planning, foresight, and inhibiting actions are under aroused in people with ADHD. Stimulants increase the neural activity in these areas to more normal levels.

Chapter 19 discussed amphetamines as potentially addictive drugs that are frequently abused. When used with medical supervision to treat ADHD, however, the drugs are safe. In addition, contrary to the notion that habitual use of these stimulants in childhood increases the chance of drug abuse or addiction in later life, the evidence suggests the opposite is true.

Up until fairly recently, the use of medication was typically stopped by adolescence; however, researchers are finding that older children and adults with ADHD can often benefit from a longer-term course of treatment. In cases where the symptoms dissipate, however, an older child may no longer need medication.

The Least You Need to Know

◆ Mental illness may be mild and allow people to carry on normal lives or may be incapacitating and require institutionalization.

◆ Scientists still debate whether mental illness has a purely physical cause, but most researchers now believe a combination of genetics and life experience is responsible for mental illness.

◆ People suffering from psychoses, such as schizophrenics, are stereotypically viewed as being incapable of functioning in the "real world." Most, however, are not confined to a hospital and can live relatively normal lives with the help of medication.

◆ Everyone gets the blues, but the feeling becomes dangerous to mental health when it becomes persistent and reaches the point of a full-blown depression. Medication and psychotherapy can help.

Part **6** Treatment (Couches, Shocks, Pills, and the Knife)

By now we hope you have some understanding of the structure and function of the brain. You have also learned about some of the things that can go wrong with the brain. Some problems occur naturally, others are the result of injury or disease, and some may be self-inflicted. As with all other aspects of the study of this incredible organ, a great deal has been learned about injury and disease, and better treatments have been developed to ameliorate and sometimes even cure problems. In this last part of the book, we explain how brain illnesses are typically diagnosed and treated. We conclude the book with a bit of prophecy about the direction of current research and the possibilities for learning more about the brain in the years to come.

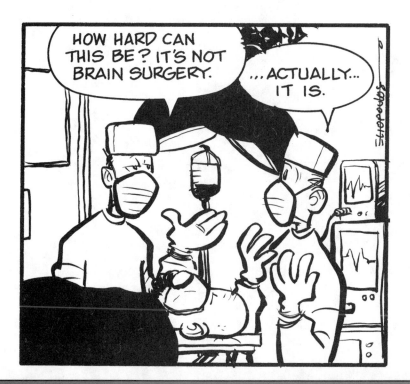

Chapter 21

Say "Ahh!"

In This Chapter

- Learning a lesson in family history
- Listening to the cerebrospinal fluid
- Testing 1, 2, 3
- Mapping the brain
- Making a nice PET

The last few chapters have described in some detail a wide variety of physical and mental illnesses connected to the brain. The question for the physician is how to determine whether a person is mentally or physically ill, whether the problem is a combination of the two, and what exactly is causing a sick person's symptoms.

In the case of physical illness, a doctor often can find a clear connection between symptoms and the disease or injury. But psychiatric and psychological problems, which often have similar symptoms, often are not as clear-cut. Scans, blood tests, and other diagnostic tools may be used to rule out physical ailments, but they don't offer any guidance with regard to determining the particular mental disorder. That *diagnosis* is usually a more subjective determination made by a mental health professional and based on the patient's history, observable behavior, and the information communicated during an examination.

Doctors typically perform a number of diagnostic tests in order to establish the diagnosis of a neurological disease. The development of imaging technology in the latter part of the twentieth century revolutionized the diagnostic capabilities of physicians. Because of the high cost of many of the scanning machines, however, most physicians and hospitals still must rely on older methods and technologies. This chapter describes the currently available tests and how they are used.

Getting to Know You

The first thing that doctors do when someone comes to see them with a problem is to get the person's personal history and perform a physical exam.

The patient's history is vitally important to the diagnostic process. The doctor wants to know when the symptoms began, how long they have lasted, and how severe they are. Have they occurred in the past? If so, did the patient receive treatment? Is there a history of similar symptoms among any members of the patient's family? Does the patient smoke, drink alcohol, or take drugs? What medication is he or she currently taking?

A physical consists of a general examination of the head, ears, eyes, nose, throat, chest, abdomen, extremities, rectal and pelvic, and it also includes an examination of the genitalia and a neurological examination. The neurological examination consists of testing the cranial nerve functions, strength in the extremities, sensation over the body, speech, and reflexes. This exam enables the doctor to quickly determine if any physical problems exist.

In addition to the history and physical exam, the doctor also gives the patient a mental status test to determine basic cognitive functions, particularly memory, orientation, and comprehension. The following questions are a sample mental status test:

Words of Wisdom

A **diagnosis** is the process of identifying an illness or injury by learning a patient's history, studying his or her symptoms, and evaluating data from laboratory and other tests.

1. What is your age?
2. What time is it (to nearest hour)?
3. I'm going to give you an address and I'd like you to remember it: 42 West Street. (The patient is asked to repeat the address to ensure it has been heard correctly and then is asked to recall the address at the end of the test.)
4. What year is it?
5. What is the name of this hospital?
6. Can you tell me who brought you to the office today?
7. What is your date of birth?
8. What year was President Kennedy assassinated?

9. What is the name of the current president?

10. Count backwards from 20 to 1.

The doctor typically scores one point for each correct response the patient gives. A person who scores 7 or 8 points or less may have a cognitive problem. However, the scale is far from perfect; 20 percent of people with a score of 7 or less are normal, and 20 percent with a score of 8 or more are cognitively impaired.

After the doctor takes the patient's history and performs physical and mental status exams, he or she may order further tests. Although skull x-rays are rarely used since the advent of CT and MRI scans, the doctor may order one if the patient has an acute head injury. Otherwise, the tests ordered are likely to be more sophisticated.

A Stab in the Back

A lumbar puncture, also known as a spinal tap, is used to diagnose several different conditions. In particular, a lumbar puncture is usually performed when there is suspicion of bleeding around the surface of the brain (called a subarachnoid hemorrhage).

For the lumbar puncture, the patient is placed on his or her side. An area of the lower lumbar region is cleaned with an antiseptic solution. The doctor feels for the depression between the spines of the third and fourth or fourth and fifth lumbar vertebra and then injects a local anesthetic. He or she then slowly advances a needle into the space around the nerves containing the cerebrospinal fluid (CSF). The pressure of the CSF is measured, and a sample of the fluid (usually 5 to 10 milliliters) is sent for laboratory analysis. The workers in the laboratory study the fluid for infection and other abnormalities by noting its color, clarity, cell count, sugar, protein, and chloride content and use stains and cultures to detect bacteria.

> **CAUTION**
>
> **Code Blue**
>
> Lumbar puncture should not be performed in a patient thought to have an intracranial mass lesion such as a brain tumor because this procedure may cause a further increase in intracranial pressure leading to permanent neurological damage or death.

The spinal tap may be unsuccessful due to arthritis, obesity, or recent spinal surgery. Should this problem occur, the doctor may obtain the CSF through a puncture at the base of the skull or in the lumbar area with x-ray control. Only an experienced neurologist, neurosurgeon, or radiologist should do these latter procedures.

Complications related to lumbar puncture are uncommon, but a headache may follow a spinal tap. The headache is usually mild and disappears within a few hours, but it may be severe and last several days. In a procedure known as blood patch application the patient's own blood is injected at the site of the lumbar puncture and usually cures the headache.

How Electrifying

Electromyography (EMG) evaluates the electrical activity in muscles through the use of needle electrodes placed in muscles. Lesions involving the nerve going to the muscle, the neuromuscular junction, and the muscle itself cause different patterns of electrical activity.

By measuring how much nerve damage has been done, doctors can better evaluate the prognosis of such conditions as Bell's palsy; that is, they can determine whether there will be no recovery, partial recovery, or complete recovery. The EMG study is important in differentiating between an abnormality of the nerve versus dysfunction of the muscle it affects.

Doctors usually perform a nerve conduction study in conjunction with the EMG. This test measures the speed of the impulse as it passes along the nerve, which helps doctors differentiate between neuropathies and between peripheral nerve problems and central nervous system disorders. Probably the most common use of these studies is to diagnose *carpal tunnel syndrome*.

Words of Wisdom

Carpal tunnel syndrome is a disorder caused by compression of a nerve in the carpal tunnel (where the nerve passes through the wrist) and is characterized by pain and numbness, tingling, and/or weakness in the hand. Since the use of computers has become more commonplace, the number of people diagnosed with this syndrome has dramatically increased. The problem is that most people hold their hands and arms in positions that compress the affected nerve while using the computer. This increase has led to the invention of various types of keyboards, cushioning devices, and other ergonomic aids designed to prevent the symptoms of carpal tunnel syndrome.

Checking Your Potentials

One way physicians can determine if a neurological problem exists is to check the electrical activity in the brain—the evoked potential—produced by an auditory, visual, or somatic (skin) stimulus. For example, a somatosensory evoked potential is the electrical activity measured in the brain with scalp electrodes produced by stimulating an area of the body, usually the arm or the leg.

Stimulation of a peripheral nerve allows recording of electrical activity over the spine and scalp. When an individual is asked to look at a pattern, electrical activity is recorded over the scalp. If a lesion exists anywhere in the visual pathway, this electrical activity will be significantly decreased or lost. The same is true for an auditory stimulus (sound).

Surgeons use these studies in the operating room to tell when a specific maneuver has caused a specific change in the evoked potential, which signals neural dysfunction. If a change has occurred, the maneuver then may be reversed, thereby minimizing or negating neurological dysfunction. In addition, these studies may be helpful in the diagnosis of multiple sclerosis when used in conjunction with the MRI.

Waves Don't Lie

Noninvasive scalp electrodes can monitor the constant electrical activity of the brain. This monitoring device, the electroencephalogram (EEG), is most helpful in the diagnosis of epilepsy. During a seizure, an individual demonstrates the abnormal activity of abrupt onset and termination. During periods of time when the individual is not experiencing a seizure, the EEG may show epileptiform (spikes on the tracing) changes.

The EEG findings aid in differentiating the various types of seizures and therefore are helpful in deciding specific drug therapy. The finding of a focal source of epileptic activity is also extremely important in the consideration of surgical therapy. The spikes seen on the EEG enable the surgeon to remove just the part of the brain causing the seizures.

However, patients who have developed a seizure disorder after a head injury may or may not have an abnormal EEG. If the EEG in this situation is abnormal for a period of time and then converts to normal, the patient still may continue to have seizures, and the normal EEG is not an indicator that the patient can be taken off anticonvulsant medications.

Gray Matter

The EEG also is useful in the diagnosis of brain death. An EEG with no electrical activity in a patient who is not hypothermic and has not had an overdose of drugs is indicative of brain death.

Take a Little Piece

Brain, nerve, muscle, and artery biopsies are used to evaluate central nervous system diseases. A neurologist may request a brain biopsy in situations where less invasive techniques have failed to establish a diagnosis. The lesion should be outside the brain stem, speech area, and motor area. Brain tumor (primary or metastatic), abscess, or degenerative disease may be diagnosed as a result of this procedure.

There are two ways to perform a biopsy. The first is an open technique in which a hole is made in the skull and the covering over the brain. A small piece of brain tissue is then removed and sent to the laboratory. The second and more common method is a stereotactic technique, which uses imaging studies such as the MRI to identify the region to be biopsied. When the region is identified, a needle is inserted through a tiny opening in the skull, and a small amount of tissue is removed through the needle and sent for laboratory analysis. The laboratory examines the tissue to determine a diagnosis so that appropriate therapy can begin.

A Bunch of "Ographies"

A number of tests are used to attempt to establish specific diagnoses. For example, the romantic-sounding transcranial doppler ultrasonography is used to detect abnormalities in intracranial blood vessels, such as constriction (vasospasm) that could cause stroke. The test is performed with a gadget that looks like a small computer mouse, which a technician moves slowly over the cranium.

Myelography involves the injection of a radiopaque dye (dye that shows up on an x-ray) at the time of lumbar puncture. The dye is then manipulated along the spinal canal, and a series of plain x-rays are taken, or the patient is placed in the CT scanner, and images of the spine are made. The MRI scan has all but replaced the myelogram, but it is still used in a few situations, such as for patients with persistent back pain for which neither the CT nor the MRI scans identified a cause of the pain.

Arteriography (angiography) requires an injection into an artery of a contrast medium (radiopaque dye) that a physician can readily see on plain x-ray. It allows the physician to see the intracranial arteries and veins in exquisite detail. It is useful in the diagnosis of intracranial aneurysms and abnormal tangles of arteries and veins called arteriovenous malformations. In regions of the country where CTs or MRIs are not available, arteriography is used to diagnose space-occupying lesions such as brain tumors. Arteriography is also used to help install detachable balloons, coils of wire, or particulate matter in blood vessels in order to block blood flow to aneurysms or arteriovenous malformations (a procedure known as embolization).

> **IQ Points**
>
> Magnetic resonance angiography has been used in recent years to study the large blood vessels in the neck and at the base of the brain. This procedure is noninvasive, but the result does not have the resolution or sensitivity of conventional angiography.

Scans Galore

Scanning procedures used in the diagnosis of neurological diseases were first used in the United States in the early 1970s. The first brain scan was the CT or CAT (computerized axial tomography) scan. The technique was developed in England by Godfrey Hounsfield at the EMI laboratory. The CT scan allowed the doctor, for the first time, to see anatomical structures within the skull using a noninvasive technique.

The first CT scanner in use in the USA was at the Massachusetts General Hospital in Boston in August 1973. Soon thereafter, a second scanner began operation at the Mayo Clinic in Rochester, Minnesota. The invention of the CT scan must be compared in significance with the introduction of the pneumoencephalogram by Walter Dandy in 1921 and cerebral angiography by Egaz Moniz in 1934. The CT scan revolutionized neurological diagnosis and led to significant changes in neurological surgical practice. The CT scan eliminated the need for drilling exploratory burr holes in the skull to look for and treat intracranial bleeding. It also allowed the neurosurgeon to more accurately approach a brain lesion, such as a tumor or abscess.

A more recently developed scan, the MRI, has enabled the physician to see brain anatomy in exquisite detail. PET, SPECT, and FMRI are all acronyms for newer types of scans that have been developed to study brain function rather than anatomy. The MRA scan enables the doctor to study the large blood vessels at the base of the brain and will, in the future, probably replace arteriography for studying all cerebral blood vessels.

CT Call Home

As you may recall from Chapter 4, the CT or CAT allows for a large number of x-rays to be taken and then processed by a computer to create a three-dimensional, anatomic image of the brain. Introduction of an intravenous contrast medium significantly increases the scan's ability to diagnose brain lesions. The scan is also very sensitive to the presence of blood so that it can be used without a contrast medium in trauma and stroke cases to identify intracranial bleeding. The CT is also useful in the diagnosis of brain tumors and degenerative diseases.

Complications from a CT are uncommon, but a patient can have a severe adverse reaction to the injected dye. The amount of radiation one is exposed to during a CT scan is also a cause for some concern for children.

A Clearer Picture

The advent of the MRI scan changed neurological and neurosurgical practice the world over. In the United States today, MRI scanners are as ubiquitous as McDonald's in many places, and most MRI scanners are in free-standing radiology centers. Unfortunately, many small hospitals across the country do not have MRI scanners, and they are still not available in many third-world countries.

Unlike a CT scan, the MRI scan does not involve radiation. In a way, the individual is within a large magnet. The magnetic field aligns the hydrogen atoms and protons within organs of the body, and then a high frequency pulse pushes them into a higher energy state. There is then a stage of relaxation. The changes in the alignment of the protons show the differences in tissue density, such as bone tissue versus muscle tissue. This information is fed into a computer, which then converts the information into an image in any desired plane of reference (for example, the sagittal, coronal, or transverse plane). The sensitivity of the MRI scan in differentiating various tissues such as bone, muscle, nerve, fat, and fluid allows for remarkable clarity in evaluating anatomical structures.

In the diagnosis of the stroke patient, the MRI scan shows an area of infarction (damage) within a matter of hours, whereas a CT scan may not show the infarction for several days. But remember the CT scan is important in the immediate (first four hours) evaluation of the stroke patient to diagnose hemorrhage.

The MRI scan is the gold standard in the diagnosis of brain tumors. Specifically, the MRI scan produces images that can be used to diagnose tumors at the base of the brain, such as acoustic neuroma, with incredible accuracy.

As we noted earlier, the MRI scan is very useful in establishing the diagnosis of multiple sclerosis. The scans can detect damage around nerve fibers in patients who will most likely develop MS, which allows doctors to put them on medication earlier and may limit the damage for years. MRI is also more sensitive in evaluating degenerative diseases such as Creutzfeld-Jakob disease. It can also be invaluable in diagnosing diseases within the spinal canal, such as benign disc herniations and tumors (benign and malignant).

> **CAUTION**
>
> **Code Blue**
>
> An MRI can't be used if a patient's body contains anything made out of metal, such as intracranial clips (used to treat aneurysms), a metallic foreign body in the eye or elsewhere, pacemakers, joint replacements (such as hip or knee replacements), and cochlear implants.

Given the coffin-like nature of the machine, individuals who are claustrophobic may have difficulty in the scanner, although newer models are more open. The open MRI has a large space so the person's body is never closer than 12 inches from the machine.

This presents an entirely different problem for these individuals than they may encounter when going through a metal detector at an airport, since the metal in the body noted by the metal detector may label the person as a possible terrorist, but the metal in the body should he enter the MRI scanner will be moved by the strength of the magnet and cause serious damage to his person.

Another version of the MRI, the functional MRI, is mainly a research tool, but recently it has been used to study the differences between degenerative diseases and schizophrenia by measuring changes in blood flow related to brain activity.

A Different Kind of PET

Positron emission tomography (PET) uses positron-emitting substances, such as radioactive glucose, to define certain areas of the brain. The PET scans the absorption of the radioactivity from outside the scalp. Brain cells use glucose as fuel, and PET works on the theory that if brain cells are more active, they will consume more of the radioactive glucose; less active brain cells consume less glucose.

A computer interprets the amount of glucose absorption and creates a color-coded brain map to display the levels of activity in different parts of the brain. Red typically represents the more active brain regions while blue indicates less active areas. The computer can examine the brain in slices, which allows for a far more detailed examination of deep brain structures than was possible with older technologies.

PET is very helpful in identifying specific areas of the kind of activity seen in epilepsy patients being considered as surgical candidates. This scan gives the surgeon the precise location within the brain from which the abnormal electrical activity originates. The PET scan is also useful in the diagnosis of degenerative diseases because it can display variations in brain metabolism by showing areas of greater or lesser oxygen consumption (glucose) and therefore brain activity.

Recent research also suggests that PET scans may enable doctors to detect Alzheimer's disease before symptoms appear. A UCLA study released in 2001, for example, found that PET detected Alzheimer's disease in 93 to 95 percent of the patients in the earliest stages of dementia. The scans also predicted whether patients would later develop the disease in nearly 90 percent of cases. This finding means that people with Alzheimer's-like symptoms may be able to find out with a high degree of certainty if they have, or will develop, the disease. For those who do have it or are predicted to get it, the PET scan allows patients to get early treatment that may help prolong their ability to function normally.

PET scans offer a number of advantages over other technologies:

♦ PET is safe.

♦ PET can replace a number of diagnostic tests with a single procedure.

♦ PET often reduces the need for invasive procedures.

♦ PET shows the progress of disease and how the body responds to treatment.

♦ PET displays all the organ systems of the body with one image.

♦ PET can sometimes diagnose a disease before other tests detect the disease.

A sometimes perplexing problem for the physician caring for a patient who has undergone surgery, chemotherapy, and radiation for a brain tumor is to tell the difference between a recurrent tumor and the death of brain tissue due to radiation (radiation necrosis). The PET scan makes this differentiation because a recurrent tumor has a high glucose uptake and therefore shows up red and yellow on the PET scan whereas dead tissue, secondary to radiation, has no glucose uptake and appears blue.

As we saw in the chapters on emotions and mental illness, PET scans are proving very useful in research into differences in brain physiology related to behavioral and cognitive tasks. The technology has been especially useful in examining brain function in schizophrenia and bipolar disorder and in the study of speech, memory, reading, and dreaming.

The Least You Need to Know

- By finding out a patient's family history, performing a physical, and using a mental status exam, a doctor can diagnose most mental and physical problems.
- A variety of diagnostic tests including lumbar puncture, electromyography, electro-encephalogram, and biopsy are used to help doctors identify the causes of their patients' symptoms.
- Modern imaging technologies, such as the CT and MRI, have revolutionized medicine and enabled doctors to make more accurate diagnoses than ever before.
- Continuing advancements in technology are allowing researchers to study the anatomy of brain structures in greater detail and to study brain function for the first time.

Chapter 22

Trick or Treatment

In This Chapter

◆ Wonder drugs

◆ Someone to talk to

◆ When in doubt, cut it out

◆ Positive thinking

Several chapters in this book have recounted all the terrible things that can afflict the brain and nervous system. The last chapter explained how doctors figure out what is wrong with the brain. This chapter describes what can be done to help.

Treatments have been mentioned throughout the recent chapters, but this chapter is more specific and detailed. The truth is that many diseases affecting the brain are incurable. Sometimes the best doctors can do is treat the symptoms and improve a sick person's quality of life. In the worst cases, a miracle is the only hope. Sadly, many people will die of illnesses of the brain. The good news is that discoveries are made every day that may lead to cures for various brain illnesses.

Even when treatments are available, a variety of barriers may make it difficult for patients to get what they need. For example, the last chapter mentioned that some of the more sophisticated imaging technologies are not universally accessible, and many people may therefore be incorrectly diagnosed. In addition, health insurance, especially under managed care, does not always cover the costs of all the procedures that may be necessary. In particular, the treatment of mental illness is often limited in terms of the number of sessions allowed by one's insurance, which may not correspond with what the doctor or therapist believes is necessary. This chapter covers the major types of treatment that are available for brain illnesses, including drugs, therapy, and surgery.

Better Living Through Chemistry

The use of medication revolutionized the mental health profession, in large part by offering an alternative to institutionalization. A number of medications have been developed for the treatment of brain injuries and disorders such as seizures, Parkinson's disease, and Alzheimer's disease.

The introduction of new drugs that are safer and more effective often encourages greater numbers of people to seek treatment for ailments. For example, the publicity surrounding the introduction of antidepressants such as fluoxetine (better known as Prozac) helped stimulate a dramatic rise in the number of people treated for depression in general and those using antidepressants in particular.

New pain-killing drugs have also been introduced, including Oxycontin, which has become one of the most widely prescribed medications for the control of pain. Unfortunately, it has also become one of the most abused of all prescription drugs.

IQ Points

The Bayer Drug Company first marketed aspirin as a pain reliever, but it was not available without a prescription until 1915.

Another problem with drugs is that they often have severe side effects. For example, Chapter 20 noted that people using older antipsychotic drugs often developed a condition called tardive dyskinesia (TD). The truth is that almost all drugs carry with them risks of side effects, some of which are serious and some of which are merely annoying.

The Placebo Effect

The term *placebo effect* is used to describe the common finding that people who participate in experiments and clinical drug trials believe that the medication they are given has helped them even if they are not treated with the medication being tested. Such studies have proven that a patient's attitude has a direct effect on his or her health, especially if he or she faces a very serious illness.

A trial to test a new medicine typically consists of two groups. One gets the medication that is being tested. The other, called the *control group*, receives something else that has no medicinal value in order to adjust for the placebo effect. Instead of a liquid medication, for example, a trial participant might be given water and told the new medication is color-less and tasteless like water. This placebo should not do anything for the control group, but many people will say the water made them feel better. If the volunteers who receive the real medication do not have significantly better outcomes (statistically speaking) than the control group, the evaluators will conclude that the medication is not effective.

According to the Surgeon General, about 30 percent of patients typically respond to a placebo in clinical trials of new antidepressants, making it very difficult to demonstrate the drugs' efficacy. A variety of explanations has been offered for this phenomenon, but the most obvious is expectation. A person in a clinical trial for a medication expects to feel better after taking the new drug, even if it's really just water.

Let's Talk

Physical problems require treatment, either physical therapy, medication, radiation, surgery, or a combination. Mental disorders, however, are primarily treated by a combination of medication and *psychotherapy*. Psychotherapy is a form of treatment for people with mental disorders and/or emotional and behavioral problems. The basic element of therapy is the one-to-one interaction between therapist and patient, with the goal of helping patients learn to cope with their problems and to change the behaviors and thoughts that make them unhappy and unable to function as they would like. Generally, psychotherapy involves either an attempt to gain insight and resolve problems by talking to a therapist or behavior modification in which patients learn how to obtain rewards through their own actions and how to unlearn the behaviors that contribute to or result from their problems.

Another way to reach the goals of psychotherapy is in a group setting. The first form of group therapy, developed in the 1920s by Austrian psychiatrist Jacob Moreno, was psycho-drama. In this method of therapy, patients act out their problems to become more aware of them as well as to better understand them. Today there are a variety of group therapy formats, and couples, families, and children now commonly visit therapists seeking help with problems related to marriage, sex, work, and family dynamics.

Words of Wisdom

Psychotherapy used to refer more specifically to the treatment of relatively severe mental illness by a psychiatrist. This definition distinguished psychotherapy from counseling provided to people with milder psychological problems by social workers and psychologists. This distinction is not as clear today. Also, the term *psychotherapy* is preferred over the term *therapy* because the latter term can refer to the general treatment of disease whereas the former specifically means the application of psychological theories of behavior.

Pick a Therapy

Sometimes there seem to be as many types of psychotherapies as there are psychiatrists and psychologists, but most types fall under one of the four major schools of psychology: psychodynamic, behavioral, cognitive, and humanistic.

Psychodynamic therapy focuses on self-understanding with an emphasis on the role of the past in shaping the present and the belief that secrets of the past are locked in the unconscious. This school of thought includes the following types of therapy:

- **Psychoanalytic**. In this stereotypical Freudian form of psychotherapy, a patient lies on a couch with the analyst out of sight and says whatever is on his or her mind (free association). The analyst tries to interpret the meaning of the patient's dreams, memories, and fantasies. The process may take years, and the cost is typically high.

- **Neo-Freudian**. A number of Freud's followers adopted modifications of his views and the goals of therapy. Erik Erikson, for example, focused on building trust and confidence for a healthy ego. He also changed the focus of conflicts from sexual to social stages and applied this stage theory to incorporate the entire lifespan. Karen Horney wanted her patients to overcome anxiety and develop realistic self-images.

- **Jungian analysis**. Carl Jung believed that Freud was too focused on the role of sex on the unconscious of the individual and didn't appreciate the importance of the experiences of all people, which has accumulated throughout evolution and is shared by everyone deep within their unconscious. This is what he called the "collective unconscious." Jung also argued that people are predominantly introverts who prefer introspection and solitary activity, or extroverts who interact with others and the environment. This form of therapy tries to help patients become more aware of their personal and collective unconscious. It also increases one's respect and knowledge of themselves and help a person function at an improved level.

- **Adlerian therapy**. Adler said people strive to reach a goal that will make them feel strong, significant within society, and complete. This form of therapy doesn't fit neatly into one of the schools, but we've included it here because it contrasts with the views of Freud and Jung who believed the components of the psyche are in conflict. Adler maintained that the conscious and unconscious work together to achieve the goals of self-improvement. Followers of Alfred Adler maintain that relationships are especially important, particularly within families. Therapists try to help patients change the way they interact with others, foster social interest and fix incorrect social values that the individual holds.

Behaviorists don't focus on understanding past events; their concern is with the actions of the present. They focus on changing or modifying that current behavior by using conditioning and other learning methods pioneered by B.F. Skinner and his followers. The following types of therapy use behavioral techniques:

- **Aversive conditioning**. This type of therapy might be seen as nearly the polar opposite of psychoanalysis because it focuses on behavior rather than what is in the mind. The behavioral therapist using this method tries to associate something unpleasant with the behavior the patient wants to change. For example, the patient might receive an electric shock when he or she begins to act in a self-destructive manner. This form of therapy is used less frequently today, and primarily for sexually deviant behavior and some drug usage.

- **Exposure and response prevention**. This behavioral technique is sometimes used to treat obsessive-compulsive disorder by preventing the patient from behaving in the way they feel compelled to do. For example, someone who compulsively washes his hands after touching something he believes has germs might be forced to touch something dirty and then the therapist would prevent him from washing his hands.

- **Relaxation**. Therapists teach their patients exercises that they can do themselves to reduce anxiety and stress.

- **Desensitization**. In this form of therapy, patients are taught to be less sensitive to situations that frighten them by having the individual either imagine or enter that situation while simultaneously relaxing to remove the connection between the situation and fear. People with phobias, for example, can be taught to relax when confronting the thing that scares them.

Cognitive psychologists focus on the idea that one's thoughts affect one's feelings and behavior. These therapists concentrate on correcting an individual's irrational beliefs, maladaptive thoughts, and illogical cognitions.

- **Rational emotive behavior therapy**. Dr. Albert Ellis first articulated REBT in 1955. It is referred to by the institute named for Ellis as "a humanistic, action-oriented approach to emotional growth which emphasizes individuals' capacity for creating their own emotions; the ability to change and overcome the past by focusing on the present; and the power to choose and implement satisfying alternatives to current patterns." The therapy holds that individuals are responsible for their own emotions and actions, that irrational thinking causes harmful emotions and actions, and that it is possible to change your way of thinking so it is more realistic and that can make life more satisfying.

- **Cognitive therapy**. Pioneered by Aaron Beck, this form of therapy deals with the present rather than the past. It focuses on our thoughts, especially unrealistic ideas that affect how we feel and behave. Therapists explore the beliefs their patients have about how to live their lives, their "rules for living," to see whether or not they are realistic and how they shape behavior. For example, a person who believes they must be the best at whatever they do is likely to find that they can't always achieve this goal and may become depressed when they fail because their rule was unrealistic. Cognitive therapy helps people become aware of their "rules" and develop realistic ones. This therapy usually involves homework in which the patients keep diaries of their thoughts, feelings, and behaviors and discuss them with their therapist.

♦ **Cognitive-behavioral therapy.** In this mix of two of the preeminent schools of thought in psychology, therapists seek to identify the thoughts that are causing a patient's problems and then may use behavioral techniques to change the patient's reactions to anxiety-provoking situations.

Humanistic approaches are especially concerned with personal growth. The past is not viewed as particularly important, and thoughts and behavior are not as significant as the patient's current feelings. The following types of therapy are considered humanistic:

♦ **Existential therapy.** Existentialism is a philosophy that emphasizes individual existence, freedom, and choice and is often associated with the belief that life is futile or meaningless. Therapists applying it in clinical settings use a variety of methods to help their patients explore anxiety, loneliness, despair, fear of death, and how to find meaning in the world.

♦ **Gestalt therapy.** This therapy originates in the work of psychotherapists Frederick and Laura Perls. Gestalt therapists attempt to make their patients more aware of themselves by focusing on the here and now of living and their relationship with other things.

♦ **Person-centered therapy.** Carl Rogers, who believed in the individual's drive for improvement and life enrichment, pioneered this type of humanistic therapy. A great deal of attention is paid to the attitude and behavior of the therapist and the way he or she interacts with the patient. The open, honest, and empathic therapist draws information out of patients and, rather than offer advice, restates the patient's insights to help them find their own solutions.

IQ Points

We couldn't find the exact origin of the word "shrink" as applied to psychiatrists. One suggestion was that the word referred to the belief that psychiatrists could reduce one's mind (shrink it) to an understandable concept. Another theory is that it is a disparaging comparison to the primitive tribal practice of shrinking heads. Unlike psychiatrists and psychologists, neurosurgeons actually can shrink the brain and do so during many surgical procedures by removing cerebrospinal fluid from the ventricular system or though a spinal tap or by using intravenous drugs such as mannitol, which causes water to be drawn out of the brain.

Today, psychotherapy is most likely to be some combination of one or more of these approaches. Like any form of treatment, psychotherapy offers no guarantees; nevertheless, it can be helpful for a wide variety of mental disorders. All the traditional techniques have been shown to be equally effective, though some work better for particular problems than others, and given the subjective nature of the process, the individual therapist can make a big difference in the comfort of the patient and the result.

Help Yourself

A therapist is not always required for someone to get help with a mental disorder. A huge industry in the United States is related to the idea of self-help. Perhaps the most successful example is Alcoholics Anonymous, which uses a 12-step program to treat alcoholism.

The Twelve Steps of Alcoholics Anonymous are as follows:

1. We admitted we were powerless over alcohol—that our lives had become unmanageable.
2. Came to believe that a Power greater than ourselves could restore us to sanity.
3. Made a decision to turn our will and our lives over to the care of God *as we understood Him.*
4. Made a searching and fearless moral inventory of ourselves.
5. Admitted to God, to ourselves, and to another human being the exact nature of our wrongs.
6. Were entirely ready to have God remove all these defects of character.
7. Humbly asked Him to remove our shortcomings.
8. Made a list of all persons we had harmed and became willing to make amends to them all.
9. Made direct amends to such people wherever possible, except when to do so would injure them or others.
10. Continued to take personal inventory and when we were wrong promptly admitted it.
11. Sought through prayer and meditation to improve our conscious contact with God, *as we understood Him,* praying only for knowledge of His will for us and the power to carry that out.
12. Having had a spiritual awakening as the result of these steps, we tried to carry this message to alcoholics and to practice these principles in all our affairs.

Off With Your Brain

Surgery and other medical interventions are rarely used to treat emotional or psychological problems, but there are some exceptions. Chapter 4 discussed the use of lobotomy and the fact that the procedure is no longer used because the damage caused by the cure was ultimately viewed as worse than the disease. Scientists have nevertheless continued to explore ways to help patients with severe mental conditions that don't respond to conventional therapies by removing parts of the brain thought to be responsible for their disorders.

One of the newer approaches to surgical treatment of mental disorders is called cingulotomy. In this procedure, a part of the limbic system, the cingulum, is destroyed. This procedure as been used to treat severe depression and bipolar disorder; however, it is effective in only about 50 percent of the cases. Although serious complications are rare, some patients have suffered paralysis because of bleeding in the brain at the time of the procedure, and in a few instances, the operation has caused seizures.

Another treatment that is being used is electroconvulsive therapy (ECT). For many of us, the only knowledge we have of ECT is what we saw in movies such as *A Beautiful Mind* and *One Flew Over the Cuckoo's Nest*. The latter, in particular, gave the lasting impression that shock therapy is used to turn people into vegetables.

Modern ECT is used almost exclusively for individuals who have severe or life-threatening mental illnesses who do not take medication or respond to it. A patient is given a general anesthetic and then a muscle relaxant. Electrodes are then placed on the head to deliver electrical impulses. The brain is stimulated for about 30 seconds, causing an epileptic-like seizure; however, there are no violent jerking motions because the medications prevent the kind of terrifying violent muscle contractions that the John Nash character suffered when he received shock therapy in *A Beautiful Mind*. Today, the treatments are typically given three times per week.

ECT does help many people with severe depression; however, its long-term impact remains controversial, with some debate as to whether it adversely affects memory.

Under the Knife

Brain surgery changed a great deal during the last 30 years as a result of the many technical advances discussed in this book; however, it is still an awesome feeling to know that one is inside someone else's brain. The neurosurgeon is acutely aware of how important it is not to make the slightest slip when operating within the brain. Not surprisingly, the neurosurgeon must have steady hands and nerves. As a result, neurosurgeons do not drink coffee prior to doing brain surgery.

You might think that the atmosphere in the operating room would be especially tense because of the importance and delicate nature of the work, but this isn't typically the case, although a certain degree of anxiety exists in the operating room prior to the beginning of the operation. Brain surgery is performed by a well-trained team consisting of the neurosurgeon, specialized nurses and technicians, and the neuroanesthetist. The operating room unit is always the same, so each member knows the other's moves, much like an athletic team, and during an operation there is the air of calm professionalism you'd expect from people who are skilled at their jobs.

Everyone knows that becoming a surgeon is intellectually challenging, but many people may not realize how physically demanding the job can be. Most brain surgery cases take from 4 to 8 hours, but it is not uncommon for a procedure to take 12 hours or longer, and some operations have lasted for a full day. Microsurgery may require the surgeon to spend long periods staring through a microscope.

Going In

The outcome of neurosurgical procedures, probably more than any other type of surgery, is dependent upon preoperative, perioperative (after the operation), and anesthetic management of the patient. Recognition of the importance of anesthesia has led to the development of the specialty of neuroanesthesia (a board-certified anesthesiologist with special training in neurological diseases). The interaction between drugs used immediately before, during, and after surgery, the manipulations of the brain during surgery, and the disease present all play a critical role in the outcome of the surgery.

Performing imaging studies (CT, MRI, and arteriography) during surgery has allowed neurosurgeons to successfully treat a number of lesions previously thought to be inoperable. Using monitoring techniques such as EEG and evoked potentials during surgery also has helped neurosurgeons produce better results with lower mortality rates.

Prep Time

As you know, the cranium is a rigid bony structure containing brain tissue, water, CSF, and blood. The delicate balance between these substances must be maintained in the preoperative period, during surgery, and in the perioperative period. This requires the expertise of a team of physicians.

Even the positioning of the patient during surgery is critical to maintain appropriate blood flow to vital areas of the brain. The control of intracranial pressure during surgery and in the perioperative period is also of paramount importance. The majority of neurosurgical procedures are done with the patient's head in a fixed metallic frame held in place with pins that attach the frame to the skull. This positioning allows the neurosurgeon to operate in delicate areas of the brain without the fear of even the slightest movement of the patient's head during the procedure.

Remember, a number of neurosurgical procedures are done with the patient awake, such as certain operations for the treatment of intractable seizures. The last thing someone wants to hear during brain surgery is the doctor saying, "Whoops!", which is why all of these procedures are done with careful preoperative evaluation and planning. All brain surgery carries a significant risk of paralysis or death and should not be taken lightly. Still, significant improvement in morbidity and mortality has been achieved over the past two decades through the use of numerous technical advances. And the future holds great promise for the neurosurgeon to further improve the rate of success of brain surgery.

The following sections describe what's involved in some of the more commonly performed neurosurgical operations. Don't worry, there's no blood involved, and you won't feel a thing.

Trauma

Operations after head trauma are the most common procedures done by neurosurgeons. The operations range from the simple elevation of a depressed (pushed-in) skull fracture to the complex combination of a severe facial fracture involving the sinuses with an intracerebral hemorrhage. The latter operation involves multiple surgeons of several specialties and can require the removal and wiring together of bone fragments, skin grafts, the removal of blood clots and dead brain tissue, control of arterial and venous bleeding, and possible repair of one or more cranial nerves.

> **Gray Matter**
>
> The more complicated procedures involving multiple injuries are typically performed in the middle of the night because this is when most severe vehicular accidents and gunshot wounds occur.

A common problem after a head injury is the development of a blood clot between the brain and it's covering, the dura, which is called a *subdural hematoma*. Drainage, and therefore removal of the hematoma, can be accomplished in most cases through the drilling of small (burr) holes in the skull.

In this operation, the surgeon makes skin incisions several inches apart in the scalp over the area of the hematoma, and then uses a drill to make openings in the skull. The surgeon then opens the dura and drains the hematoma with a suction device. The holes in the skull may be covered with plastic covers or left open with only the skin over them. The skin is then sutured. In many cases, this is a life-saving procedure.

Case Study Number One

An 18-year-old high school senior was brought to the emergency room unconscious after a single vehicle accident at 3 A.M. He was comatose, with the strong smell of alcohol on his breath. He had a compound frontal skull fracture, and cerebrospinal fluid was leaking from a wound above his right eye. He was rushed to surgery, where a blood clot was found in his right frontal lobe. As the clot was removed, the gear shift knob of his car was found embedded in his brain and was carefully removed. The dura, which had been torn at the time of the incident, was repaired with a graft taken from a cadaver (cadaver dural grafts used to be kept in a freezer for emergency use, but in recent years, they have been replaced by artificial dura). The patient awakened within an hour of the operation, but he subsequently developed an infection within the brain (from the dirty gear shift knob), leading to the formation of a brain abscess that surgeons later removed. Months later, a plastic plate was put in his head to replace skull fragments that had been removed in the first operation. A year later, he appeared normal in all respects.

Hydrocephalus

Hydrocephalus is a condition involving excess fluid collection and enlargement of the cerebral ventricles. It is seen during the first three months of life and is often diagnosed at the time of birth.

To treat hydrocephalus, surgeons make a small opening in the skull above and behind the right ear, and then place a hollow plastic tube through the brain into the fluid-filled ventricle. The surgeon then creates a tunnel is under the skin from the small skin incision in the scalp to a small skin incision in the abdomen and places a hollow plastic tube in this tunnel with the end in the abdominal cavity. The surgeon then attaches a one-way valve to the tube entering the ventricle; the surgeon attaches the other end of the valve to the tube entering the subcutaneous tunnel that ends in the abdominal cavity. This procedure allows the cerebrospinal fluid to be "shunted" to the abdominal cavity, precluding the development of increased intracranial pressure and abnormal head growth.

Craniofacial Anomalies

This unfortunate childhood condition is characterized by an abnormal configuration of the skull and face. Some of these conditions are caused by a premature closure of skull sutures (the interlocking connections or interdigitations of the various bones of the skull that allow for normal skull growth).

To obtain the best surgical result, these children should be treated in the neonatal period. In most cases surgery is performed to improve appearance; however, certain conditions if untreated can lead to significant neurological deficit and/or mental retardation, especially in cases where there is associated hydrocephalus.

Surgery may involve several incisions to open prematurely closed sutures, for the removal of bone, and for the advancement of other bones, particularly around the bony socket of the eye (the orbit). In recent years, these operations have been performed by teams of surgeons that include a plastic surgeon, ENT (ear, nose, and throat) surgeon, and a neurosurgeon. Results of this surgery have continued to improve over the last decade as more plastic surgeons have taken an interest in the problem. Early treatment can protect the child from severe facial deformity and ridicule later in life.

Brain Tumors

Surgery for benign brain tumor (pituitary tumor, acoustic neuroma, meningioma) begins with the premise that the entire tumor will be removed. Before surgery, doctors locate the tumor using CT or MRI studies. The location of the tumor dictates how the surgery is performed.

In the case of a malignant tumor (an astrocytoma or glioblastoma) or a metastatic tumor, removing the tumor is usually impossible because finger-like projections of the tumor extend into vital parts of the brain. However, recent advances in surgical techniques, including the use of a robotic arm to perform intricate manipulations of instruments and intraoperative MRI, have allowed neurosurgeons to remove more of the malignant tumor with fewer deaths than ever before.

Pituitary Tumors

Pituitary tumors are approached by making an incision above the upper teeth inside the mouth and then extending into the nasal cavity. Using the operating microscope, the neurosurgeon can reach behind the nose into the sphenoid sinus and through a thin shelf of bone into the cranial cavity to see the pituitary tumor. The tumor is then removed by suctioning, and a few stitches close the small hidden incision above the teeth. This procedure means the tumor can be removed without opening the skull at the top of the head.

Gray Matter

The pituitary gland is found within a portion of the skull called the sella turcica, so-called because it looks like a Turkish saddle.

Case Study Number Two

A 29-year-old female noted amenorrhea (stoppage of menstrual periods) for a period of six months and an inability to become pregnant. When she developed a thin milky discharge from the breast (galactorrhea), she saw her gynecologist. A careful examination revealed no pelvic abnormalities, but laboratory tests indicated hormonal abnormalities. An MRI scan showed the presence of a pituitary tumor. The tumor was removed, and she started having normal menstrual periods. Eighteen months later, she became pregnant.

Acoustic Neuroma

Surgery for acoustic neuroma usually requires the expertise of the neurootologist as well as the neurosurgeon. These benign tumors sit at the base of the brain in the angle between the cerebellum and the brain stem. The goal of the surgery is to preserve facial nerve function and hearing if the patient has any demonstrated on preoperative audiology tests. If the patient has lost hearing, the approach to the tumor usually is to make an incision behind the ear and then drill with a diamond drill through the mastoid air cells to gain access to the tumor. Small tumors can be readily removed through this approach.

Large tumors, or any size tumor that could affect hearing, should be removed through what is called a posterior fossa approach. In this procedure, the surgeon makes an incision behind and below the mastoid air cells. The surgeon then makes an opening in the skull over the cerebellum and approaches the tumor from the side and below. This technique allows for preservation of facial function and hearing in many cases. The covering over the cerebellum is repaired, and the bone that was removed may or may not be replaced. The muscles of the neck that were cut to gain access to the skull are sutured, as is the skin.

Case Study Number Three

A 44-year-old male noted a gradual decrease in hearing in his right ear. He noticed that he now held the telephone to his left ear, something he had not done previously. He had no other complaints. His family physician referred him to an otologist (a physician who specializes in hearing problems) who performed a test for hearing (audiogram) that revealed a 50 percent hearing loss in the right ear and normal hearing in the left ear. An MRI scan revealed a small tumor (less than two centimeters) in the right cerebellopontine angle that was consistent with a benign acoustic neuroma. Working together, a neurosurgeon and a neurootologist completely removed the tumor. The patient's hearing was preserved, and mild facial weakness that was present immediately after surgery cleared completely in three months.

Meningioma

In the case of a meningioma in the frontal part of the brain, the surgeon makes an incision behind the hairline and pulls back the scalp to access the skull. Using a high-speed drill (air driven or electric), the surgeon makes a small opening in the skull. The surgeon then uses a power-driven saw to remove a portion of the skull very carefully so as not to tear the covering over the brain (dura). The size of the plate of the skull that is removed depends on the size of the underlying tumor. The covering over the brain is then opened, exposing the surface of the brain. During the remainder of the operation, the surgeon typically uses a microscope to see better.

The tumor is removed using ultrasound suctioning or a laser, with careful attention paid to not damaging adherent brain tissue or important blood vessels. The surgeon then sews the covering over the brain back into place and replaces the skull plate. The operation ends with the skin being sutured and a sterile bandage being applied.

Case Study Number Four

A 37-year-old male went to his doctor because he'd been having increasingly painful headaches for the past six months and because his wife had noticed a change in his personality. On physical examination he appeared normal. His neurological examination revealed a slight slurring of speech and a minimal weakness of his right arm and leg. An MRI scan revealed a large tumor in the left frontal area of the brain. A cerebral arteriogram showed the arteries supplying the tumor and the typical appearance of a benign meningioma. The patient's tumor was successfully and completely removed, and he had an uncomplicated recovery.

Metastatic Tumors

Surgical treatment of metastatic tumors presents special problems. In most cases, if imaging studies reveal the presence of multiple tumors, surgical excision is not indicated. These patients will be treated with radiation and chemotherapy. If a solitary lesion is found in the area of the brain controlling speech or motor function, it may be removed surgically. In every case an attempt should be made to find the site of the primary tumor (that is the organ of origin of the brain metastasis) so that this area can be treated.

Case Study Number Five

A 56-year-old male noted gradually increasing weakness in his left arm over a period of three weeks. He had a grand mal seizure and was taken to the emergency room where he was found to have paralysis of his left arm. He was awake and alert after recovering from the seizure. An MRI scan revealed a tumor in the motor area of his right cerebral hemisphere. He was started on a course of steroids (to decrease brain swelling) and anticonvulsant medication (to prevent further seizures) and scheduled for surgery the following day.

At the time of surgery, a large tumor was found 1 centimeter deep to the surface of the brain. About 90 percent of the tumor was removed, but the rest was too deep in vital areas of the brain. A biopsy taken during surgery was reported as "highly malignant, compatible with glioblastoma multiforme" (the most malignant form of brain tumor). The patient's initial postoperative course was uneventful, and the strength in his arm improved. He was subsequently treated with a course of radiation therapy and chemotherapy, but unfortunately the malignancy recurred 14 months later, and he died.

Words of Wisdom

A **noninvasive** technique does not require a skin incision or an opening into the skull.

A Radioactive Knife

Many of the tumors described previously may be treated with the gamma knife. This *noninvasive* method delivers pinpoint radiation to the tumor with little or no effect on adjacent normal brain tissue.

You might then ask why not treat all brain tumors with the gamma knife? The answer is that they do not all respond to this type of therapy.

Cerebral Aneurysms and Arteriovenous Malformations

Brain surgery may also be performed to treat an aneurysm (a bubble on a weakened area of an artery). In most cases of aneurysms, the goal of the operation is to prevent blood from flowing into the bubble. This can be done by excluding the aneurysm from the artery by placing a metal clip (that looks like a clothespin) along the base of the bubble. An arteriovenous malformation or AVM, a tangle of arteries and veins, may be removed using the same technique of entry into the skull as is used for the removal of a brain tumor. In addition, aneurysms and AVMs may be treated by embolization (the insertion of something into the blood vessel to block blood flow) with tiny coils of wire, detachable balloons, or a type of glue. In some cases AVMs can be treated with the Gamma knife, obviating the need for an open surgical procedure.

To accomplish this embolization, a specially trained neurosurgeon or neuroradiologist places a plastic tube in an artery in the thigh and then, under x-ray control, threads it up the aorta in the abdomen and then the carotid artery in the neck and finally into the artery in the brain where the abnormality (aneurysm or AVM) is located.

In a very small number of patients this procedure may have significant complications, including stroke with paralysis or death. This procedure should only be done in centers with specially trained personnel and where specialists are available to treat the complications of the procedure should any occur.

> **CAUTION**
>
> ### Code Blue
>
> If you blow up an old inner tube and it develops a weakened area that looks like a bubble, the tire has just what an aneurysm looks like. Aneurysms are not removed very often because doing so would leave a hole in the blood vessel that would cause the patient to bleed to death.

Case Study Number Six

A 24-year-old female was brought to the emergency room by her fiancé after she complained of the onset of an excruciating headache during sexual intercourse. The headache persisted unabated. She also complained of a stiff neck and being somewhat drowsy. An MRI scan was normal. A spinal tap revealed the presence of blood in the spinal fluid. An emergency arteriogram revealed the presence of an aneurysm of the intracerebral portion of the internal carotid artery. Doctors felt that this aneurysm had ruptured, which caused the headache and the blood in the spinal fluid. The following morning the patient was taken to the operating room where the aneurysm was treated by placing a metal clip at it's base, thereby isolating it from the cerebral circulation. Because blood could no longer enter the aneurysm, there was no chance that it could bleed again. The patient made an uneventful recovery from what is generally considered one of the most high-risk surgical procedures and was informed that it was perfectly safe to resume sexual relations.

Stroke

Surgical treatment of stroke can be divided into three categories:

◆ Prevention

◆ Emergency blood clot removal

◆ Revascularization (providing blood flow to an area that has been deprived by the stroke)

In many cases of stroke, the cause is an embolus (clot) that originates from an atherosclerotic plaque in the carotid artery in the neck. This clot causes blockage of an artery to a particular area of the brain, resulting in death of the neurons in that area (infarct).

Several surgeries may prevent the stroke from occurring. The carotid endarterectomy involves making an incision in the mid-portion of the neck between the chin and shoulder. The neck muscles are moved to one side, and the carotid artery is identified by its pulsation. Blood flow through the artery is temporarily halted by the use of a tourniquet around the artery. The artery is then opened and the atherosclerotic plaque is removed, thereby relieving the blockage of the artery. Think of a garden hose with a rock in it. The rock interferes with the flow of water. When the rock is removed, the flow is improved. After the artery is cleared, the neck muscles are allowed to resume their normal position, and the skin is sutured closed.

In the presence of an acute stroke, when an emergency CT scan reveals the presence of a large intracerebral hematoma (blood clot), surgery may be advised if the patient has signs of increased intracranial pressure. The blood clot is removed using the same techniques that are used to remove brain tumors.

Revascularization is a procedure that restores blood flow to an area devoid of blood flow. This can be accomplished by connecting the extracranial artery (the superficial temporal artery in front of the ear) to a branch of the middle cerebral artery on the surface of the brain. This procedure may be used in patients who have TIAs or RINDs (small strokes) and do not have demonstrable narrowing (stenosis) of the carotid artery due to an atherosclerotic plaque.

Treatments vs. Cures

Everyone who is ill wants to get better and thanks to advances in medical science, it is often possible to ameliorate symptoms and to sometimes conquer diseases. Many injuries and disorders of the brain that once were the cause of permanent disability or death can now be treated. Still, no one should be confused about the difference between treatments and cures. Treatments can ameliorate some symptoms of a disease while a cure can protect you from a disease or reverse its effects. Not even the best intentions, the most skilled physician, or the most sophisticated technology can cure everything. Tragically, there are types of brain tumors and illnesses such as Creutzfeldt-Jakob disease that are fatal. There is always hope, and new medical advances are always being developed, but we don't want anyone to mistakenly believe that the treatments we've described in this chapter will necessarily cure all brain-related ills.

One key to good health is to have the support of friends and family. This support system can be helpful in providing informal counseling and encouraging you to get treatment and to stick with it when symptoms of illness appear. Emotional support is very important to anyone who has a physical or mental disorder.

People with extremely positive attitudes toward their illnesses sometimes recover or live longer and enjoy a better quality of life. Religious faith and the determination to get better has also been shown to help individuals with serious illnesses, including brain disorders.

The Least You Need to Know

- Medical science can't cure all ills, but every day new advances are made toward curing more diseases and ameliorating their symptoms.
- A wide variety of psychological approaches have been developed to treat mental disorders. Despite their differences in theory and application, they are all equally effective.
- For many centuries brain surgery was a death sentence, but modern techniques have made even complex brain surgery a relatively safe procedure.
- It is difficult to explain scientifically, but a positive outlook, a determination to get better, and a strong religious faith can make a difference in a person's fight against an illness.

Chapter 23

We're Getting Smarter

In This Chapter

- ◆ Staying humble
- ◆ It's in the genes
- ◆ Ethical questions
- ◆ Send in the clones
- ◆ Stranger than fiction

We're coming down to the wire. Hopefully, you haven't developed a headache, become depressed, or taken any illegal drugs in the course of making it this far.

Every day researchers learn more about the workings of the brain. In the latter half of the twentieth century, researchers learned more than they had known in all the previous centuries put together. With adequate funding for research, a commitment to scientific inquiry, and a lot of luck, this exponential growth in knowledge will continue.

The United States, in particular, has made a commitment to pursue brain research. As you recall from Chapter 4, the 1990s were the "decade of the brain." In addition, the Human Brain Project is a federal research initiative that involves more than a dozen government organizations that are encouraging and supporting behavioral research, basic and clinical neuroscience, and computer-based resources designed to assist in the study of the brain and its functions. This chapter covers some of the latest research on the brain.

Mistaken Science

The ancient views about the brain and the ideas that were developed during the early stages of brain research may seem naive or even absurd to us now. However, we should be humble about our own notions because they will one day be considered the ancient ones and our ideas will no doubt be viewed with equal incredulity.

We don't even have to look that far back into history to see our mistakes. For example, earlier we discussed how scientists in the early twentieth century incorrectly thought that a variety of negative traits, such as a tendency toward criminal behavior, were hereditary, and policymakers used this scientific finding to justify compulsory sterilization of habitual criminals.

Genius for Genes

In February 2001, researchers announced the first comprehensive interpretation and analysis of the *human genome*. The project has given hope to researchers who believe that a biological basis underlies most, if not all, mental illnesses and physical disorders of the brain. The hope is that it will now be possible to identify specific genes that cause degenerative diseases, such as Alzheimer's, as well as mental disorders such as schizophrenia and depression.

The study of the human genome offers a number of intriguing possibilities for the treatment of disease. For example, gene therapy offers the potential of altering genes that cause disease, but the problem is that scientists understand the function of very few of the estimated 30,000 to 35,000 genes, and some disorders may involve multiple genes.

> **Words of Wisdom**
>
> The **human genome** represents all of the DNA, including the genes, in each of us. Genes carry information for making all the proteins in the body, and these proteins determine what we look like, how we respond to infection, and, to some degree, how we behave.

One offshoot of the genetic research is pharmacogenomics, a mouthful that means the study of how an individual's genetic inheritance affects the body's response to drugs. The notion is that drugs could be customized to fit an individual's genetic makeup and that safer and more powerful medications can be developed. Though gene therapy has received a lot of media attention, especially after a patient in one clinical trial died, the technique is still in its infancy and is currently in an experimental phase with only a small number of research projects involving humans.

Ethical Dilemmas of Modern Science

As physicians and researchers develop more advanced technologies, they face difficult ethical dilemmas. If you saw the film *Jurassic Park*, you may recall the mathematician, played by Jeff Goldblum, who made the point that the scientists were so focused on seeing if they *could* develop the technology to bring dinosaurs back to life that they never asked themselves whether they *should*. The moral in that story, like that of *Frankenstein* many years before, was clearly no.

Today, research is often overseen by a variety of review boards and typically scrutinized for ethical quandaries. A whole academic discipline in medical ethics evolved in the latter half of the twentieth century, and now the ethical component of every scientific revelation is immediately questioned in public forums.

To take a recent case, an announcement was made in February 2002 that doctors in Chicago had used sophisticated genetic tests on human eggs to help a woman give birth to a child they believed was free of the family's curse of early Alzheimer's disease. The news immediately set off a firestorm of protest from people who believed this testing was another step toward creating designer babies who would have the traits the parents wanted and would not have traits that were either dangerous, as in the case of the Alzheimer's gene, or just undesirable, say, short stature or a large nose.

Code Blue

The genetic screening test for early Alzheimer's can't be used for the more common form of the disease because no one gene has been identified as being responsible for the disorder. However, scientists have discovered a genetic abnormality associated with the rare form of Alzheimer's that typically hits people in their 30s and 40s.

Despite the safeguards that have been put in place and the public scrutiny that follows research, there are always scientists who are prepared to push the envelope. Some may have different visions or be less concerned with ethical issues or simply come to different conclusions about them. Someone is always willing to go where no one has gone before.

Who Benefits?

In addition to questions of whether research should be pursued, ethical questions also arise over how to ration and use new technologies. We mentioned earlier the idea of developing a "smart pill" and noted that this pill could have a variety of attendant ethical issues. Should everyone get them? Should you be allowed to take them right before tests?

Cost is invariably an issue. If a smart pill were developed, but it was exorbitantly priced, should we allow it to be sold on a normal market where only the wealthy could afford it and the poor would have no access to a medication that could dramatically change their lives?

Creating Monsters

In addition to improving our lives, science can make our lives more precarious. Nuclear power, for example, has many benefits, but it has also created new dangers from radioactive waste (not to mention nuclear war). Researchers looking for cures to diseases may inadvertently discover new diseases that have no cure. Government-sponsored research here and abroad on biological weapons has produced agents, some specifically targeting the nervous system, that are designed to make people sick and/or kill them.

Sometimes new treatments simply don't work or are worse than the disease, despite the best of intentions. For example, a promising vaccine was discovered to halt and, in some cases, reverse the effects of an Alzheimer's-like condition in mice, and it was being used in clinical trials on humans around the world. However, the drug was withdrawn from the tests after reports of brain inflammation in several human volunteers.

From the Stem

One potentially revolutionary and highly controversial area of research involves the use of stem cells. A stem cell has the potential to develop into any type of cell in the body and can multiply to provide a constant source of new cells. Stem cells are produced during the embryonic period of development. These unspecialized cells produce hundreds of different kinds of neurons. Many researchers believe that stem cells have the potential of allowing the creation of new organs or cells to replace or renew diseased hearts, livers, and perhaps even brains.

Most studies currently use stem cells on rats. In one study reported in early 2002, for example, stem cells were injected into the brain of rats affected by the symptoms of Parkinson's disease. Once in place, the stem cells converted to neurons that make dopamine, the neurotransmitter that is lacking in patients with Parkinson's. As always, one has to keep in mind it is always a leap from research on animals to research on humans, but this study is one of many promising examples of the potential of stem cell research.

Gray Matter

For a long time, no one knew that stem cells were active in the adult brain. We now know that they are in the hippocampus.

The use of stem cells is especially controversial because embryos have the potential to become people. In the United States federally funded research on human embryonic stem cells is limited because producing such cells requires the death of human embryos. President Bush approved some research using the cells, but he limited it to the approximately 60 cell lines that already exist from embryos that were previously destroyed.

A vigorous debate is raging over stem cells. Supporters of research argue that the embryos were going to be destroyed anyway and that research from their cells holds the potential for important medical advances. Critics say stem cell research is wrong because it destroys human life. Most stem cells have come from embryos left over from fertility treatments or from abortions, which has enmeshed the issue in the controversy over abortion. The situation has been further complicated by the creation of embryos specifically for the purpose of stem cell research.

Hello Dolly!

In 1997, a sheep born at a laboratory in Scotland startled the scientific world and provoked an ongoing debate about the possibilities for curing diseases and the dangers of tampering with nature. The sheep, named Dolly, was born to a surrogate mother and had no father. She was not the product of the usual union of sperm and egg; instead, Dolly's genetic material had come from cultured cells taken from a mature ewe, making her a clone (a genetic replica) of the donor sheep. Scientists had cloned other animals before her, but Dolly was the first mammal to be cloned from an adult, and she raised the theoretical possibility of one day cloning a human.

The issue of cloning is one of the most controversial in science and in society. We can now clone lots of living organisms, including, most recently, a cat. However, even before the announcement of the first successful cloning of a sheep, questions have been raised about the wisdom of pursuing this research and how far it should be allowed to go. For many people, the red line is any effort to clone human beings, and the U.S. government has taken legal steps to prevent this research. Even so, scientists are certain to continue to pursue the goal of cloning a human, either in the United States in a private laboratory that may not be subject to government restrictions or in a foreign country where fewer limits are placed on research.

IQ Points

In early 2002, the announcement of the first successful cloning of a cat was met with a combination of excitement and disdain. Researchers at Texas A&M University successfully cloned a kitten as part of research funded by a wealthy private donor who hopes they will one day successfully clone his dog Missy. Cloning a cat is easier than cloning a dog (because scientists understand a feline's reproductive cycle better), so this was a preliminary step. The technique opens the possibility for a huge business for cloning pets, an idea that upsets some people who believe it is dangerous and adds to the overpopulation of pets in the United States. Though the sponsor and researchers deny an interest in cloning humans, such cloning remains an implication of the technology.

Cloning offers many possibilities for medical science. For example, modifying the genes in animals may make it possible to use their organs for human transplants. Animals could also be created with genetic defects to allow for research into cures and treatments (which is not a popular option among animal rights activists). Cells might also be cloned for use in therapies for illnesses such as Parkinson's disease. Cloning might lead to the development of herds of cattle that lack the gene that gives rise to Mad Cow disease.

For now, however, the potential is far greater than the reality. Even if all ethical concerns were set aside, scientists don't know enough about the human genome to customize genes to make improvements in people. In addition, the cloning process is by no means perfected. The success rate is very low. For example, it took 276 tries to clone Dolly and 87 tries before a feline clone survived.

For those concerned about the implications of cloning, the news that Dolly developed arthritis was a red flag. In January 2001, the researchers who had cloned the sheep said that not only was the five-and-a-half year old suffering from this disease, but signs had also been found that she was aging prematurely. Based on the information available today, however, it is not known whether these conditions are related to the cloning procedure or are simply coincidental.

Deep Blue

An entire field of research known as artificial intelligence is devoted to creating machines that have the qualities of the human brain, including intelligence, creativity, and perceptiveness. Scientists have developed robots and computers with increasing levels of sophistication. Perhaps the most celebrated of these was the IBM computer named Deep Blue that defeated chess world champion Gary Kasparov. Of course, humans programmed the computer.

> **IQ Points**
>
> The idea of combining humans and machines has long been the subject of science fiction. In the Steven Spielberg film *AI*, a time is imagined when realistic robots are a part of society and it becomes possible to create one with human emotions. In *The Matrix*, the humans have jacks in the back of their skulls that allow them to be connected to a computer.

Perhaps an even more interesting field of study is neuroprosthetics, which involves research related to restoring neural function using artificial means. For example, implants can be used to create alternative pathways for signals to travel to and from the brain to compensate for damaged pathways. Cochlear implants that aid the hearing impaired are an example of this kind of device, as are electrodes implanted in the brains of Parkinson's patients. Other researchers are hoping to give sight to the blind by implanting electrodes in the visual cortex.

One study that attracted a lot of media attention in March 2002 involved the implantation of electrodes in the brains of monkeys. These electrodes tapped into neurons in the motor cortex, and the researchers were able to convert the neural pulses to electrical signals. The monkeys were trained to use a mouse to move a cursor on a computer screen. After the electrodes were inserted, the computer mice were turned off, and the scientists found the monkeys were able to move the cursor just by thinking about it. When the monkeys realized they didn't need to use their hands, they used the mouse less and simply moved the cursor with their thoughts. Among the implications of this research, if it can be replicated and later duplicated in humans, would be to enable people who are paralyzed to use their thoughts to control electronic instruments such as computers.

A New Brain?

Although doctors now can transplant lungs, livers, hearts, and pancreases, no one has yet figured out how to transplant a brain, but that is not to say it can never be done. The challenge to make all the connections to make anything work is enormous because of the intricacy of the relationship between the brain, brain stem, and spinal cord, not to mention the billions of neuronal links.

Researchers have been successful, however, in transplanting neurons. The initial tests of these neurons were on stroke patients, but these neurons may also be beneficial in the treatment of brain injury, Parkinson's disease, and spinal cord injury. The brain has a number of barriers, however, that inhibit nerve cell regeneration or transplantation that scientists are trying to overcome.

Stem cells have also been transplanted into the brain. The hope is that replacing missing or faulty genes with new cells may make it possible to replace cells destroyed by diseases or strokes, or stimulate regeneration of nerves after brain and spinal cord injuries.

IQ Points
In the 1970s, Professor Robert White of Cleveland transplanted the head of a monkey onto the body of another monkey. The controversial operation was criticized by some as a mad exercise that offered no medical benefit, but White maintained the procedure was the logical next step in research that had also been controversial when first used on other organs, especially the heart. White argued that people with paralysis and other disabilities might ultimately benefit from such transplants. One problem White and subsequent researchers couldn't overcome was making connections beyond some of the major arteries and veins to allow the monkey with the new head to move its body.

The cost of health care has a dramatic impact on the treatment of brain-related injuries and disorders. Without going into the larger debate about insurance and managed care, the reality is that many of the procedures we have described in this book and some of the advanced technology and medication is very expensive. For example, some of the new antipsychotic drugs cost significantly more than the older ones; therefore, doctors are often forced to prescribe the cheaper medication first and are allowed to use the newer ones only if the older ones fail.

From Our Brains to Yours

This is the end of our crash course in understanding the brain. If you are like us, you will feel inspired to learn more, even if it means referring to writers who do not have the distinction of being authors of a *Complete Idiot's Guide*.

Don't feel bad if you don't remember everything or didn't understand it all. Whether you have a terrific memory or a poor one, the good news is you can always refer back to this book.

We hope you have enjoyed learning about the brain as much as we've enjoyed writing about it. The brain is a fascinating and extraordinary organ that still holds many mysteries. We look forward to the day when they will be unlocked, and then we'll revise this book and share all the answers with you. In the meantime, keep using that brain of yours; it'll serve you well.

The Least You Need to Know

- We have learned a great deal about the brain, especially in the last few decades, but if history is our guide, future generations are likely to discover that we've made mistakes.

- We are unlocking the secrets of human genes, which can help us determine the cause of diseases and develop treatments and possible cures.

- Technology is allowing us to do many things that raise ethical questions as to whether we should do things just because we can.

- Although we are far from the technology needed for a brain transplant, researchers are transplanting nerve cells.

Appendix A

Glossary

acupuncture   An ancient system of healing developed in the Far East that involves the insertion of fine needles into specific locations of the body.

alchemy   A practice in which practitioners claimed the ability to mysteriously transform ordinary materials into something special. Some alchemists believed they could change base metal into gold, cure diseases, and indefinitely prolong life.

anterior   Toward the front.

antiseptics   Substances that prevent or arrest the growth of microorganisms.

apgar score   Refers to the evaluation of newborns based on five characteristics—color, heart rate, response to stimulation of the sole of the foot, muscle tone, and respiration. Infants are rated 0, 1, or 2 for each characteristic with a total of 10 being a perfect score.

aromatherapy   Oils from plants are used to treat a variety of disorders. The fragrant substances are sometimes massaged into the skin and other times inhaled.

asepsis   A method by which harmful organisms are killed so they are never present in the operating room. For example, all furniture, instruments, and linens in a surgery room are treated with antiseptics.

aura   Subjective phenomenon or sensation marking the onset of a seizure. This may be auditory (you hear strange sounds), visual (you see strange things), motor (you have strange movements), or gastrointestinal (you have a sensation of abdominal pain).

biopsy   The removal and examination of tissue, cells, or fluids from the body.

capillary   Derived from the French or Latin word for hair. Capillaries are microscopic blood vessels that form a network of tiny tubes throughout the body, connecting the smallest arteries and veins. Capillaries have very thin walls composed of a single layer of cells that distribute oxygen and nutrients from the blood into the body tissues and absorb waste and carbon dioxide.

carpal tunnel syndrome   A disorder caused by compression of a nerve in the carpal tunnel (where the nerve passes through the wrist). It is characterized by pain and numbing, tingling, and/or weakness in the hand.

cochlea   From the Latin for "snail," this is a part of the inner ear that is coiled like a snail shell and is responsible for hearing.

conditioned reflex   Learned behavior that is distinguishable from an innate reflex, which is automatic, such as pulling your hand away from a flame.

cortex   From the Latin word for the "bark" (of a tree). This name is appropriate because the cortex is a sheet of tissue that makes up the outer layer of the brain.

crack   Cocaine processed in a ready-to-use form for smoking. The term refers to the crackling sound heard when the mixture is heated.

decibels (dB)   A measure of sound intensity.

dementia   Condition that involves the deterioration of mental functions, such as memory, concentration, and judgment, resulting from an organic disease or a disorder of the brain. It is sometimes accompanied by emotional disturbance and personality changes and, typically, interferes with a person's daily functioning.

diagnosis   Process of identifying an illness or injury by learning a patient's history, studying his or her symptoms and evaluating data from laboratory and other tests.

diseases   Conditions with detectable physical changes.

disorders   Typically have unknown pathology and are associated with an impairment of functioning.

DNA   Short for deoxyribonucleic acid. The acronym is shorthand for describing the nucleic acids in cells that form the shape of a double helix and are the molecular basis of heredity.

drug addiction   The compulsive use of drugs despite harmful consequences.

Electra complex   Freudian theory that girls are envious of their father's penis and want to possess it so strongly that they dream of bearing his children. This "penis-envy" leads to resentment toward the mother, who the girls believe caused their castration.

evoked potential   The electrical activity produced by an auditory, visual, or somatic stimulus. A somatosensory evoked potential is the electrical activity produced by stimulating an area of the body, usually the arm or the leg.

fight-or-flight reaction   The body's natural response to stress in which blood pressure, heart rate, and muscle tension are among the functions adjusted to prepare to confront or evade a threatening situation.

fissure   A slit or groove between body parts.

flashbulb memories   In 1977, Roger Brown and James Kulick suggested that people could so vividly recall what they were doing at the time of John F. Kennedy's assassination because shocking events such as these activate a special brain mechanism they referred to as "Now Print." The Now Print mechanism acts like a camera's flashbulb, freezing in our mind whatever happens at the moment when we learn of the shocking event. These memories are referred to as "flashbulb memories."

foramen magnum   The opening in the base of the skull through which the spinal cord exits the skull and goes into the neck.

forebrain   Front part of the brain that includes the cerebrum, thalamus, and hypothalamus.

hallucinogens (also psychedelics)   Drugs that affect a person's perceptions, sensations, thinking, self-awareness, and emotions.

hertz (Hz)   A measure of sound frequency, which is the number of cycles of vibrations per second.

hindbrain   The lower region of the brain stem, comprising the pons and medulla.

histology   The study of the structure of tissues.

homeopathy   Treatment developed by a German doctor named Samuel Hahnemann (1755–1843), who believed that chemicals that caused diseases could act as cures if used in extremely small doses.

hominids   The definition of our human family whose characteristics include a large, highly developed brain, our upright position, and our manner of movement. We are also distinguished from other animal families by our use and construction of tools.

human genome   All the DNA, including the genes, in each of us.

hypnosis   Term coined by James Braid around the middle of the nineteenth century. It is also referred to as mesmerism after Franz Anton Mesmer, and describes a condition that resembles sleep but is induced by another person.

idiot savant   An intellectually disabled person with unusual ability in a particular field, such as art, music, or mathematics. The condition was first named in 1887 by Dr. J. Langdon Down. He used the word "idiot" because that was the accepted classification for the mentally retarded (IQ below 25) at the time. A "savant" (from the French savoir, meaning "to know") is a learned person. The word "idiot" is no longer used to describe this condition because it is considered derogatory.

Jerusalem syndrome   Many people make pilgrimages to the holy city of Jerusalem. Some people become so intoxicated by the spirituality of the place that they begin to act erratically and some develop temporary or long-term delusions of grandeur, such as the belief that they are the messiah or that they can speak directly to God or Jesus. This occurs so often that psychiatrists recognized it as an illness.

LASIK   The acronym for laser-assisted in situ keratomileusis. In LASIK surgery, precise and controlled removal of corneal tissue by a special laser reshapes the cornea and thus changes its focusing power. Before deciding on this procedure, a person should consult with one or more eye surgeons.

lateral   The direction away from the midline.

linguistics   The scientific study of language. One branch of this field, neurolinguistics, specifically examines how language is processed and represented in the brain.

method of Loci   Dates back to ancient Greece when Greek orators would use this method to help them memorize speeches. One story relates that the Greek orator Simonides of Ceos was at a banquet to give a speech and stepped outside for a moment. At that moment, the building collapsed, and everyone inside was killed, and their bodies were mangled beyond recognition. Simonides identified the bodies of the guests based on where he remembered them sitting or standing before he left the building.

midbrain   The uppermost part of the brain stem.

natural selection   A process that results in the survival of plants or animals that are best suited to their environment and leads to the perpetuation of genetic qualities best adapted to that environment.

neuroses   Disorders that have no apparent organic cause. Basically all the mental illnesses were once referred to as neuroses, but the word has fallen out of favor and is now rarely used by mental health professionals.

neurotransmitter   A substance that transmits nerve impulses across a synapse.

noninvasive   A technique that does not require a skin incision or an opening into the skull.

Oedipus complex   Freudian notion that holds that boys have an unconscious attraction to their mothers and hostility toward their fathers.

peripheral nervous system   The part of the nervous system, which includes the autonomic system, that is anatomically outside the brain and spinal cord, although physiologically it is intimately related to the brain and spinal cord. This system has 36 pairs of peripheral nerves; 31 pairs are spinal nerves that enter the central nervous system below the neck, and 5 pairs are cranial nerves that connect directly to the brain.

plaques   Small, disk-shaped growths. More specifically, they are lesions of brain tissue that consist of abnormal clusters of dead and dying nerve cells.

prodigies   People with exceptional talents that are usually displayed at a young age.

proprioreceptors   Sensors within the body that detect stimuli from the muscles and tendons and help us orient our limbs and bodies.

psychotherapy   The treatment of mental illness by means of therapy that applies psychological theories of behavior.

pyramidal system   This system gets its name from the pyramid-shaped neural bundles in the area of the medulla that most of the pyramidal tracts cross.

Renaissance   Literally "rebirth." The Europeans believed they were rediscovering Greek and Roman culture after centuries of intellectual and cultural decline and returning civilization to greatness.

shingles   A skin rash caused by the virus causing chicken pox.

shunt   Device used in a surgical procedure that diverts (shunts) cerebrospinal fluid from the ventricular system into the abdominal cavity where it can be absorbed.

skeletal muscles   Muscles that are attached to bones and look striped under a microscope.

smooth muscles   Muscles that appear smooth when looked at under a microscope and are associated with internal organs.

stress   Hans Seyle defined stress as "the nonspecific response of the body to any demand made upon it." He called the "demands" stressors.

trephining   Ancient practice of cutting holes in the skull, which may have been intended to release from the brain or mind the evil spirits and demons believed to cause mental and physical illnesses.

ventromedial   The lower part in the center or midline.

Appendix B

Food for the Brain: Resources

If your brain is still hungry for more information about how it works, here are some books and websites that are sure to sate it!

Books

Barmeier, Jim. *The Brain*. San Diego: Lucent Books, 1996.

Beckelman, Laurie. *The Human Body*. Weldon Owen Inc., 1999.

DeMoss, Robert. *Brain Waves Through Time*. NY: Plenum Trade, 1999.

Gilman, Sid and Sarah Winans Newman. *Clinical Neuroanatomy and Physiology Manter and Gatz*. Philadelphia: F. A. Davis Co., 1992.

Gray, Henry. *Gray's Anatomy*. Philadelphia: Lea and Febiger, 1942.

Greenfield, Susan. *The Human Brain*. NY: Basic Books, 1997.

Johnston, Joni. *The Complete Idiot's Guide to Psychology*. Indianapolis: Pearson Education, Inc., 2000.

Kotulak, Ronald. *Inside the Brain*. Kansas City: Andrews and McMeel, 1996.

Madras, Bertha K. "Changing Your Mind: Drugs in the Brain." CD-ROM, Cambridge: President and Fellows of Harvard College, 1996.

Netter, Frank. *The CIBA Collection of Medical Illustrations*. NJ: CIBA Pharmaceutical Company, 1983.

Physicians Desk Reference. Montvale: Medical Economics Co., 1999.

Ratey, John. *A User's Guide to the Brain*. NY: Pantheon, 2000.

Restak, Richard. *The Secret Life of the Brain*. The Dana Press and the Joseph Henry Press, 2001.

Simon, Roger P., and Michael J. Aminoff and David A. Greenberg. *Clinical Neurology*. Appleton and Lang, 1999.

Singer, Charles. *A Short History of Medicine*. Oxford University Press, 1962.

Websites

A Brief Tour of the Brain   http://www.uib.no/med/avd/miapr/arvid/UiB50/syracus/biology.htm

American Stroke Association   http://www.StrokeAssociation.org

Antiqua Medicina: From Homer to Vesalius   http://www.med.virginia.edu/hs-library/historical/antiqua/anthome.html

BrainConnection.com   http://brainconnection.com/

Brain Facts and Figures   http://faculty.washington.edu/chudler/facts.html

Brain Surgery Information Center   http://www.brain-surgery.com/

Brain: The World Inside Your Head   http://www.pfizer.com/brain/index.html

Centers for Disease Control   http://www.cdc.gov/

Creativity and the Brain by Dr. Arnold Schiebel, PBS TeacherSource   http://www.pbs.org/teachersource/scienceline/archives/sept99/sept99.shtm

Cyber Museum of Neurosurgery   http://www.neurosurgery.org/cybermuseum/

Drug Enforcement Agency   http://www.usdoj.gov/dea/

Epilepsy Foundation   http://www.efa.org

History of Neuroscience   http://faculty.washington.edu/chudler/hist.html

Human Anatomy Online   http://www.innerbody.com/htm/body.html

Human Genome Project   http://www.ornl.gov/hgmis/

International Dyslexia Foundation http://www.interdys.org/index.jsp

MedicineNet.com http://medicinenet.com/Script/Main/hp.asp

Multiple Sclerosis Association of America http://www.msaa.com/

National Institute of Mental Health http://www.nimh.nih.gov/

National Organization on Fetal Alcohol Syndrome http://www.nofas.org/

National Stroke Association http://www.stroke.org

Neuroscience for Kids http://faculty.washington.edu/chudler/neurok.html

Neurosurgery://On Call http://www.neurosurgery.org/

Office of the Surgeon General http://www.surgeongeneral.gov/sgoffice.htm

Scientific American.com http://www.sciam.com/

Secret Life of the Brain http://www.pbs.org/wnet/brain/

The BrainWeb & Brain Information http://www.dana.org/brainweb/

The Harvard Brain http://hcs.harvard.edu/~husn/BRAIN/

The Parkinson's Web http://spauldingrehab.mgh.harvard.edu/parkinsonsweb/Main/PDmain.html

Society for Neuroscience Brain Backgrounders http://www.sfn.org/content/Publications/BrainBackgrounders/index.html

Survey of Neurobiology by David L. Atkins http://gwis2.circ.gwu.edu/~atkins/Neuroweb/Neuronode.html

Vision, Navigation, Mobility and the Brain http://isd.saginaw.k12.mi.us/~mobility/visbrain.htm

Web of Addictions http://www.well.com/user/woa/facts.htm

WebMD http://www.webmd.com/

Wisconsin Medical Society http://www.wisconsinmedicalsociety.org/savant/default.cfm

World Health Organization http://www.who.int/home-page/

Appendix C

Test Your IQ

This test is an untimed verbal test with 25 questions. It has a bias toward the American English language. Taking this test without any aids such as a dictionary will give you the most accurate score.

1. Under the school administration's _____ rule, the students didn't dare misbehave.

   A. draconian

   B. democratic

   C. benevolent

   D. lax

2. Under the Khmer Rouge's _____ regime, millions of Cambodians were _____.

   A. glorious/prosperous

   B. despotic/murdered

   C. bloodless/relocated

   D. democratic/enfranchised

   E. incompetent/well-fed

3. The following sentence makes sense if the word "beuss" is understood to mean the same as the word "play": Even though some of the beussers had never beussed before, they all did their best to beuss as hard as they could.

   A. True

   B. False

4. Can the word FLOUNDER be spelled using letters from the word WONDERFUL?

   A. Yes

   B. No

5. If the word MAD is written under the word CAP and the word CAT is written under MAD, then the word MAT is formed diagonally.

   A. True

   B. False

6. Can the words MAILMAN, COMMINGLE, and ALMANAC each be spelled using letters from the word MEGALOMANIAC?

   A. Yes

   B. No

7. gnaw : psyche :: hour : ___?

   A. final

   B. zest

   C. wrote

   D. nadir

   E. three

8. Borglum : Founding Fathers :: Roebling : _____ ?

   A. Russian Revolution

   B. French Resistance

   C. U.S. Civil War

   D. Hoover Dam

   E. Brooklyn Bridge

   F. Berlin Airlift

9. EXPIATE : ATONE ::

   A. explicate : confuse

   B. expropriate : dispossess

   C. expatiate : restrain

   D. expiscate : speculate

   E. expostulate : condone

   F. expurgate : vomit

10. PLUTOCRACY : WEALTHY ::

    A. stratocracy : military

    B. autocracy : many

    C. timocracy : anarchists

    D. ochlocracy : police

    E. ergatocracy : elite

11. K : M :: 6076 : ___?

    A. 6076

    B. 5280

    C. 5820

    D. 7456

    E. 6074

    F. 5764

12. II : 10 :: III : ___?

    A. 10

    B. 3

    C. 20

    D. 11

    E. 100

    F. 1

13. (ONE : (ONE : FOUR)) : (3 : (9 : 16)) :: (TWO : (FOUR : NINE)) : ___?

    A. (2 : (4 : 16))

    B. (5 : (36 : 49))

    C. (4 : (25 : 36))

    D. (4 : (16 : 25))

    E. (6 : (36 : 64))

    F. (4 : (16 : 36))

14. (holo : (ounce : gram)) : (kilo : (mile : meter)) :: (contra : (band : music)) : ___?

    A. (school : (foot : yard))

    B. (ear : (beat : drum))

    C. (band : (length : width))

    D. (loco : (motion : sensor))

    E. (light : (measurement : weight))

15. Some lollipops are red. Everything red has stripes. Therefore, which of the following must be true?

    A. All lollipops have stripes

    B. Everything red is a lollipop

    C. Some lollipops have stripes

    D. All lollipops are red

    E. None of the above

16. Insert the word missing from the brackets.

    monitor (tote) computer

    economic (_ _ _ _) agenda

17. Anthracite : cirrus :: down : ___?

    A. goose

    B. duck

    C. low

    D. up

    E. sad

18. Blow : snake :: strike: ___?

    A. push

    B. bit

    C. venom

    D. hurt

    E. assistant

19. Allison is shorter than Eric, but she is taller than Nate. Gillian is shorter than Eric, and she is shorter than Nate. Chris is taller than Gillian, but shorter than Eric. If the statements above are true, one can validly conclude that Mike is shorter than Allison if it is true that

    A. Allison is equal in height to Chris

    B. Chris is equal in height to Mike

    C. Mike is taller than Gillian, but shorter than Chris

    D. Mike is shorter than Eric, but taller than Nate

    E. Chris is taller than Mike, but shorter than Nate

20. (cata : (tonic : gin)) is to (epi : (center : front)) as (para : (dictatorship : military)) is to ____?

    A. (neo : (para : chute))

    B. (con : (niption : fit))

    C. (tele : (call : phone))

    D. (for : (psycho : ward))

    E. (demi : (god : man))

    F. (dis : (locate : pair))

21. Choose the word that is most similar to LUGUBRIOUS:

    A. tired

    B. mournful

    C. hungry

    D. gelatinous

    E. stately

22. Paradox : catch 22 :: bourgeois : ___?

    A. psittacine

    B. plebian

    C. socialist

    D. babbitt

    E. shebolith

23. Cross : mark :: polka : ___?

    A. pattern

    B. dance

    C. song

    D. sing

    E. German

24. Slip : mess :: beef : ___?

    A. flab

    B. alb

    C. step

    D. steak

    E. rumor

25. Troop : company :: bison : ____?

    A. division

    B. tigers

    C. unit

    D. platoon

    E. parrots

Now check your answers against this answer key:

1. A	14. D
2. B	15. C
3. A	16. MIND
4. A	17. D
5. B	18. A
6. A	19. E
7. C	20. C
8. E	21. B
9. B	22. D
10. A	23. B
11. B	24. C
12. D	25. E
13. D	

To score your test, give yourself 3 points for each correct answer and add the total to 90. For example, if you get 10 questions correct, you add 30 points to 90, giving you an IQ of 120. The IQ scores correspond to those in the scale we presented earlier and duplicate below. Only 6.1 percent of the people who take the test score in the top 5 percent. Note that this particular test does not allow for people to score below normal, so it is primarily designed to determine whether you have above average intelligence.

140 and over	Genius or near genius
120–140	Very superior intelligence
110–120	Superior intelligence
90–110	Normal or average intelligence

This test is supplied courtesy of the International High IQ Society and the questions are the property of the Society. If you score 126 or above on the test, you qualify for membership in the International High IQ Society. To further test yourself, and learn more about membership, visit the Society's website at www.highiqsociety.org.

Index

autism, 264
autoimmune diseases, 230
autonomic nervous system,
    129-130
    fight-or-flight response,
        131-133
    parasympathetic nervous sys-
        tem, 135
    proprioception, 138
    reflexes, 136-137
    sexual behavior, 138-139
    temperature control, 130
        enteric nervous system,
            130
        sympathetic and para-
            sympathetic nervous
            systems, 130-135
aversive conditioning, 283
Avicenna, 20
AVM (arteriovenous malforma-
    tion), 293
awareness levels, reticular for-
    mation, 72
axons, 79

B

babbling (language develop-
    ment), 99
bacterial meningitis, 210
balance, 141-142
    cerebellar function, proprio-
        ception, 69-70
    cranial nerve eight, 90-91
    extrapyramidal systems,
        146-147
    kinesthetic sense, kinesthet-
        ics, 142
    labyrinthine sense, 142-144
    vestibular system, 69
balance of water, thirst, 120
basal ganglia, 62
    amygdala, 72
Bedlam, 253
behavioral therapies, 282
    aversive conditioning, 283
    desensitization, 283

exposure and response pre-
    vention, 283
relaxation, 283
Bell's Palsy, 32
    cranial nerve seven, 90
Berger, Hans, 50
beta waves (brain waves), 122
Binet, Alfred, 43-44, 185
Bini, Lucino, electroshock, 50
Binnig, Gerd Karl, STM (scan-
    ning tunneling microscope), 51
binocular vision, 108
biological influence (emotions)
    versus environmental,
        170-172
biological needs
    eating, hypothalamic con-
        trol, 119-120
    sexual behavior, 126
        hypothalamus' role, 126
        sexual orientation, 127
    sleep, 120-126
biopsies, 214, 273-274
bipolar disorder, 263-264
blackouts, 163
bladder control, propriorecep-
    tion, 138
blind spot, 107
blink reflex, 137-138
blood circulation, 25
blood clots, 205
    emergency blood clot
        removal, 294
blood pressure regulation
    cranial nerve nine, 91
    medulla oblongata, 71
blood supply to the brain, 4
blood-brain barrier, 60
    Ehrlich, Paul, 38
bloodless surgical approach
    (Dandy), 48
bloodletting, 22
bodily-kinesthetic intelligence,
    183
body of neurons, 79
    axons, 79
    dendrites, 79

body hair, temperature control,
    117
body smart. *See* bodily-
    kinesthetic intelligence
body versus mind, 23
    Descartes, René, 24-25
Bois-Reymond, Emil, 30-31
bones of the skull (cranium), 58
bony labyrinth, 143
    motion detection, 144
bovine spongiform
    encephalopathy. *See* BSE
bow-wow theory, 98
bowel control, propriorecep-
    tion, 138
brain attacks (stroke), 216-217
    hemorrhagic, 217-218
    ischemic, 217
    treatment, 218-219
brain death, 219
    organ transplantation, 220
brain stem, 71
    medulla oblongata, 71
    pons, 71
    reticular formation, 72
Brain Tumor Registry, 47
brain waves (EEG), 122
breathing regulation
    cranial nerve ten, 91
    medulla oblongata, 71
Broca's area, 63, 101
BSE (bovine spongiform
    encephalopathy), 233

C

caduceus, 17
caesarean section, 19
calcium, 82
capillaries, 60
carbamazepine (Tegretol), 264
cardiac muscle, 146-147
cardiovascular function, reticu-
    lar formation, 72
carotid Doppler studies, 218
carotid endarterectomy, 218
carotid sinus, 91

F

S